全国电子信息类优秀教材

大学计算机系列教材

JSP大学实用教程

（第2版）

◆ 耿祥义　张跃平　编著

电子工业出版社
Publishing House of Electronics Industry
北京·BEIJING

内 容 简 介

JSP 是一种动态网页技术标准，可以建立安全、跨平台的先进动态网站。

本书详细讲解了 JSP 的重要内容，特别注重结合实例讲解一些难点和关键技术。全书共分 11 章，内容包括：JSP 简介、JSP 页面、JSP 标记、内置对象、JSP 与 JavaBean、文件操作、数据库操作、Java Servlet、MVC 设计模式，以及应用实例——网上书城。本书所有知识都结合具体实例进行介绍，力求详略得当，突出 JSP 在开发 Web 动态网站方面的强大功能，使读者快速掌握和运用 JSP 的编程技巧。

本书可以作为高等院校计算机及相关专业的教材，也适合自学者及网站开发人员参考使用。

未经许可，不得以任何方式复制或抄袭本书之部分或全部内容。
版权所有，侵权必究。

图书在版编目(CIP)数据

JSP 大学实用教程 / 耿祥义，张跃平编著. —2 版. —北京：电子工业出版社，2012.1
大学计算机规划教材
ISBN 978-7-121-14365-6

Ⅰ. ① J… Ⅱ. ① 耿… ② 张… Ⅲ. ① JAVA 语言－网页制作工具－高等学校－教材
Ⅳ. ①TP312 ②TP393.092

中国版本图书馆 CIP 数据核字（2011）第 166208 号

策划编辑：章海涛
责任编辑：章海涛　　　　　特约编辑：何　雄/张　玉
印　　刷：北京虎彩文化传播有限公司
装　　订：北京虎彩文化传播有限公司
出版发行：电子工业出版社
　　　　　北京市海淀区万寿路 173 信箱　邮编　100036
开　　本：787×1092　1/16　印张：18.5　字数：474 千字
版　　次：2007 年 5 月第 1 版
　　　　　2012 年 1 月第 2 版
印　　次：2021 年 7 月第 17 次印刷
定　　价：36.00 元

凡所购买电子工业出版社图书有缺损问题，请向购买书店调换。若书店售缺，请与本社发行部联系，联系及邮购电话：(010) 88254888。

质量投诉请发邮件至 zlts@phei.com.cn，盗版侵权举报请发邮件至 dbqq@phei.com.cn。
服务热线：(010) 88258888。

第 2 版前言

本书是《JSP 大学实用教程》的第 2 版，更新了部分例子和内容，增加了一个实训内容，即第 11 章（实训二：网上书城）。本次修订特别注重教材的可读性和实用性，许多例题都经过精心的考虑，既能帮助理解知识，同时又具有启发性。

本书在 2012 年被中国电子教育学会评为"全国电子信息类优秀教材"。

全书共分 11 章，分别讲解了 JSP 页面、JSP 标记、内置对象、JSP 与 JavaBean、文件操作、数据库操作、Java Servlet、MVC 设计模式等重要内容。

第 1 章介绍 JSP 重要性，对 Tomcat 服务器的安装与配置进行了详细介绍。

第 2 章讲解 JSP 页面的基本构成以及 JSP 页面的运行原理。

第 3 章讲述常用的 JSP 标记，特别对 Tag 标记给予详细的讲解，使读者认识到代码复用的重要性。

第 4 章讲解内置对象，特别强调了这些内置对象在 JSP 应用开发中的重要性，结合实例使读者掌握内置对象的用法。

第 5 章是 JSP 技术中很重要的内容，即怎样使用 JavaBean 分离数据的显示和处理，这一章给出了许多有一定应用价值的例子。

第 6 章主要讲解怎样使用 Java 中的输入/输出流实现文件的读写操作，在实例上特别强调怎样使用 JavaBean 实现文件的读写操作。

第 7 章涉及的内容是数据库，也是 Web 应用开发的非常重要的一部分内容，特别介绍了各种数据库的连接方式。

第 8 章讲解 Java Servlet，对 Servlet 对象的运行原理给予了细致的讲解，许多例子都是大多数 Web 开发中经常使用的模块。

第 9 章对 Java Servlet 在 MVC 开发模式中的地位给予了重点介绍，并按着 MVC 模式给出了易于理解 MVC 设计模式的例子。

第 10 章是一个完整的网站，完全按着 MVC 模式开发设计，其目的是使读者掌握一般 Web 应用中常用基本模块的开发方法。

第 11 章是一个完整的网站，完全按着 JSP+Tag 模式开发设计，其目的是使读者掌握一般 Web 应用中常用基本模块的开发方法。

希望本书能对读者学习 JSP 有所帮助，并请读者批评指正。

本书为任课教师提供配套的教学资源（包含电子教案、例题源代码、部分习题解答），需要者可登录到华信教育资源网（http://www.hxedu.com.cn），注册之后进行下载。

读者反馈：unicode@phei.com.cn，xygeng0629@sina.com。

作 者

目 录

第1章 JSP 简介 1
   1.1 什么是 JSP 1
   1.2 Tomcat 服务器的安装与配置 1
   1.3 测试 JSP 页面 4
   1.4 设置 Web 服务目录 5
   1.5 设置端口号 7
   习题 1 7

第2章 JSP 页面 8
   2.1 JSP 页面的基本结构 8
   2.2 JSP 的运行原理 9
   2.3 JSP 页面的成员变量和方法 12
   2.4 JSP 页面中的 Java 程序片 13
   2.5 JSP 页面中的 Java 表达式 15
   2.6 JSP 中的注释 16
   2.7 在 JSP 页面中使用 HTML 标记 17
   习题 2 23

第3章 JSP 标记 25
   3.1 指令标记 page 25
   3.2 指令标记 include 28
   3.3 动作标记 include 29
   3.4 动作标记 param 31
   3.5 动作标记 forward 31
   3.6 动作标记 useBean 33
   3.7 Tag 文件与 Tag 标记 33
   习题 3 39

第4章 内置对象 40
   4.1 request 对象 40
   4.2 response 对象 46
   4.3 session 对象 51
   4.4 out 对象 60
   4.5 application 对象 61
   习题 4 64

第5章 JSP 与 JavaBean 65

5.1 编写和使用 JavaBean ··· 66
  5.1.1 编写 bean ··· 66
  5.1.2 使用 bean ··· 67
5.2 获取和修改 bean 的属性 ·· 71
  5.2.1 动作标记 getProperty ·· 71
  5.2.2 动作标记 setProperty ·· 73
5.3 bean 的辅助类 ·· 77
5.4 举例 ·· 79
  5.4.1 三角形 bean ·· 79
  5.4.2 四则运算 bean ··· 80
  5.4.3 猜数字 bean ·· 82
  5.4.4 时间 bean ··· 84
  5.4.5 日历 bean ··· 87
  5.4.6 播放幻灯片 bean ··· 90
习题 5 ··· 91

第 6 章 JSP 中的文件操作 ·· 93

6.1 获取文件信息 ··· 94
6.2 创建、删除 Web 服务目录 ·· 96
6.3 读写文件 ··· 98
  6.3.1 读写文件的常用流 ·· 98
  6.3.2 读取文件 ·· 100
  6.3.3 按行读取 ·· 103
  6.3.4 写文件 ··· 105
6.4 标准化考试 ··· 106
6.5 文件上传 ·· 111
6.6 文件下载 ·· 113
习题 6 ·· 114

第 7 章 在 JSP 中使用数据库 ··· 115

7.1 SQL Server 2000 数据库管理系统 ·· 116
7.2 JDBC ·· 117
7.3 数据库连接的常用方式 ·· 118
  7.3.1 JDBC-ODBC 桥接器 ·· 118
  7.3.2 使用纯 Java 数据库驱动程序 ·· 122
7.4 查询操作 ·· 125
  7.4.1 顺序查询 ·· 126
  7.4.2 随机查询 ·· 132
  7.4.3 条件查询 ·· 136
  7.4.4 排序查询 ·· 141

  7.4.5 模糊查询 ··· 143
 7.5 更新、添加与删除操作 ·· 145
 7.6 使用预处理语句 ·· 153
  7.6.1 预处理语句优点 ··· 153
  7.6.2 使用通配符 ··· 155
 7.7 基于 CachedRowSet 分页显示记录 ··· 158
 7.8 常见数据库的连接 ·· 163
  7.8.1 连接 Oracle 数据库 ··· 163
  7.8.2 连接 MySql 数据库 ··· 164
习题 7 ·· 170

第 8 章 Java Servlet 基础 ·· 171

 8.1 Servlet 对象的创建与使用 ·· 172
  8.1.1 HttpServlet 类 ·· 172
  8.1.2 部署 Servlet ·· 172
  8.1.3 运行 Servlet ·· 174
 8.2 Servlet 工作原理 ·· 175
 8.3 通过 JSP 页面调用 Servlet ·· 176
 8.4 Servlet 的共享变量 ·· 179
 8.5 doGet()方法和 doPost()方法 ··· 180
 8.6 重定向与转发 ·· 183
 8.7 会话管理 ·· 186
  8.7.1 获取用户的会话 ··· 186
  8.7.2 猜数字 ·· 188
习题 8 ·· 191

第 9 章 基于 Servlet 的 MVC 模式 ·· 192

 9.1 MVC 模式介绍 ··· 193
 9.2 JSP 中的 MVC 模式 ··· 193
 9.3 模型的生命周期与视图更新 ·· 194
 9.4 MVC 模式的简单实例 ··· 196
  9.4.1 计算三角形的面积 ··· 196
  9.4.2 四则运算 ··· 199
  9.4.3 读取文件 ··· 203
  9.4.4 查询数据库 ··· 206
习题 9 ·· 212

第 10 章 实训一：会员管理系统 ·· 213

 10.1 系统模块构成 ·· 213
 10.2 数据库设计 ··· 214

10.3	系统管理	214
10.4	会员注册	217
10.5	会员登录	223
10.6	上传照片	227
10.7	浏览会员	232
10.8	修改密码	240
10.9	修改注册信息	244
10.10	退出登录	250

第 11 章 实训二：网上书城　　252

11.1	系统主要模块	252
11.2	数据库设计与连接	253
11.3	系统管理	255
11.4	会员注册	257
11.5	会员登录	259
11.6	浏览图书信息	262
11.7	查询图书	265
11.8	查询购物车	268
11.9	订单预览	271
11.10	确认订单	273
11.11	查询订单	276
11.12	查看图书摘要	278
11.13	修改密码	280
11.14	修改注册信息	282
11.15	退出登录	286

第 1 章

JSP 简介

> **本章导读**
> ✿ 知识点：了解 JSP 的来历以及在开发动态网站上的优势。掌握 Tomcat 服务器的安装与配置。
> ✿ 重点：Tomcat 服务器的安装与配置。
> ✿ 难点：学习怎样设置 Web 服务目录。
> ✿ 关键实践：上机编写、保存、运行一个简单的 JSP 页面。

1.1 什么是 JSP

Java 语言以不依赖于平台、面向对象、安全等优良特性成为网络程序设计语言中的佼佼者。目前，许多与 Java 有关的技术得到了广泛的应用和认可，JSP（Java Server Pages）技术就是其中之一。JSP 是基于 Java 语言的一种 Web 应用开发技术，可以建立安全、跨平台的先进动态网站。许多 Web 网站都使用了 JSP 技术。利用 JSP 技术创建的 Web 应用程序可以实现动态页面与静态页面分离，便于 Web 应用程序的扩展和维护。由于 JSP 是基于 Java 语言的 Web 技术，相对其他 Web 技术，JSP 具有脱离硬件平台束缚、编译后运行等优点，已成为 Internet 上的主流 Web 技术之一。

1.2 Tomcat 服务器的安装与配置

网络通信中最常见的模式是 B/S 模式，即需要获取信息的用户使用浏览器向某个服务器发出请求，服务器进行必要的处理后，将有关信息发送给服务器。在 B/S 模式中，服务器上必须有所谓的 Web 服务软件和 Web 应用程序，Web 服务软件负责处理客户对 Web 应用程序的请求，并负责运行管理 Web 应用程序，以满足客户对信息的请求。因此，学习 JSP 需要安装一个支持 JSP 的 Web 服务软件，这样的软件也称为 JSP 引擎。将安装 JSP 引擎的计算机称为一个支持 JSP 的 Web 服务器。支持 JSP 的 Web 服务器负责运行 JSP，并将运行结果返回给用户，有关 JSP 的运行原理将在本书 2.2 节中讲解。

目前，比较常用的 JSP 引擎包括 Tomcat、JRun 和 Resin，以 Tomcat 的使用最广泛。Tomcat 软件是一个免费的开源 JSP 引擎，也称为 Tomcat 服务器。Tomcat 服务器由 Apache 和 Sun 公司共同开发而成，可以登录到 http://jakarta.Apache.org/tomcat 免费下载 Tomcat 6.0。登录之后，先在 Download 页面中选择 apache-tomcat-6.0.13，然后在 Binary Distributions 的 Core 页面中选择 Zip 或 Windows Service Installer 即可。如果选择 Zip，将下载 apache-tomcat-6.0.13.zip 文件；如果选择 Windows Service Installer，将下载 apache-tomcat-6.0.13.exe 文件。

本书的重点在于学习 JSP 本身，不涉及比较各种 JSP 引擎的优缺点，因此选择 Tomcat 服务器来学习 JSP。本节主要介绍 Windows 2000/XP 操作系统下 Tomcat 服务器的安装配置。

1．安装 JDK

安装 Tomcat 之前，首先安装 JDK，这里安装 Sun 公司的 JDK 1.6。假设 JDK 的安装目录是 D:\Jdk1.6。安装 JDK 之后需要进行几个环境变量的设置。对于 Windows XP/2000，右键单击"我的电脑"，在弹出的快捷菜单中选择"属性"命令，弹出"系统特性"对话框，再单击其中的"高级选项"，然后单击按钮"环境变量"，分别添加如下系统环境变量：

　　变量名：Java_Home　　　　　变量值：D:\jdk1.6
　　变量名：Classpath　　　　　变量值：D:\jdk1.6\jre\lib\rt.jar;.;
　　变量名：Path　　　　　　　　变量值：D:\jdk1.6\bin

如果曾经设置过环境变量 Java_Home、Classpath 和 Path，可单击该变量进行编辑操作，将需要的值加入即可，如图 1.1、图 1.2 和图 1.3 所示。

2．安装与启动 Tomcat 服务器

（1）apache-tomcat-6.0.13.zip 的安装

将下载的 apache-tomcat-6.0.13.zip 解压到磁盘某个分区，如解压到 D:\，解压缩后将出现如图 1.4 所示的目录结构。

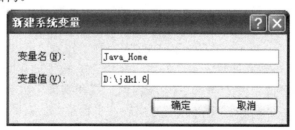

图 1.1　设置 Java_Home

图 1.2　设置 Classpath

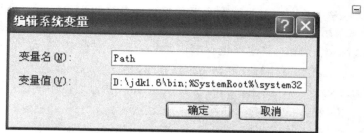

图 1.3 编辑 Path

图 1.4 Tomcat 服务器的目录结构

执行 Tomcat 安装根目录中 bin 文件夹中的 startup.bat 或 tomcat6.exe 来启动 Tomcat 服务器。执行 startup.bat，启动的 Tomcat 服务器会占用一个 MS-DOS 窗口（如图 1.5 所示），如果关闭当前 MS-DOS 窗口，将关闭 Tomcat 服务器。建议使用 startup.bat 启动 Tomcat 服务器，以确保 Tomcat 服务器使用的是 Java_Home 环境变量设置的 JDK。

图 1.5 启动 Tomcat 服务器

（2）测试 Tomcat 服务器

在浏览器的地址栏中输入"http://localhost:8080"或"http://127.0.0.1:8080"，会出现如图 1.6 所示的 Tomcat 服务器的测试页面。

图 1.6 测试 Tomcat 服务器

注意：Tomcat 服务器默认占用 8080 端口，如果 Tomcat 所使用的端口已被占用，Tomcat 服务器将无法启动，有关端口号的配置稍后讲解。

（3）文件 apache-tomcat-6.0.13.exe 的安装

文件 apache-tomcat-6.0.13.exe 是针对 Windows 的 Tomcat 服务器的，安装后形成的目录结构与 apache-tomcat-6.0.13.zip 完全相同。

双击下载的 apache-tomcat-6.0.13.exe 文件，将出现"安装向导"界面，单击【Next】按钮，接受授权协议后，将出现选择安装方式界面，从中选择"Normal"、"Minimun"、"Custom"或"Full"等安装方式。比如，选择安装方式为"Full"后单击【Next】按钮，将出现选择安装目录界面,从中可以给出安装 Tomcat 的目录,如输入安装目录为 E:\Tomcat6.0。在选择安装目录时,最好不要使用该界面给出的默认目录,以方便今后使用 Tomcat。在选择安装目录界面中选择安装目录后,单击【Next】按钮,将出现设置端口号和管理密码设置界面。Tomcat 服务器必须占用一个端口号,以便与其他网络程序相区分。该界面提供的 Tomcat 服务器占用的默认端口号是 8080,在安装 Tomcat 时,可以在该界面设置 Tomcat 服务器所占用的端口号,但最好不要使用 1～1024 之间的端口号,以免与其他网络程序发生冲突。另外,在该界面中可以输入管理密码,以便以后管理 Tomcat 服务器。在该界面中,将 Tomcat 服务器的端口设置为默认端口 8080,管理密码设置为 123456。

对于 Windows 2000/NT/XP,在安装 Tomcat 成功时,安装程序会提示用户选择"即刻启动 Tomcat"或"稍后启动 Tomcat"。如果没有选择"即刻启动 Tomcat",可以通过"开始"→"所有程序"→"Apache tomcat 6.0"→"start Tomcat"启动 Tomcat,也可直接执行 Tomcat 安装目录 bin 文件夹中的 startup.bat 或 tomcat6.exe 来启动 Tomcat,如 E:\Tomcat6.0\bin\tomcat6.exe。注意:若 Tomcat 所使用的端口已被占用,Tomcat 将无法启动。

1.3 测试 JSP 页面

Tomcat 正确启动后,需要测试它是否是一个 JSP 引擎,因为 Tomcat 默认是一个 HTML 引擎。在 1.2 节中,在浏览器的地址栏中输入"http://127.0.0.1:8080"后,Tomcat 将 index.html 的超文本文件发送给浏览器,该超文本文件中并没有 JSP 技术所涉及的内容,因此可以看见 Tomcat 测试页,但并不能保证 Tomcat 的安装就完全正确无误。

我们将使用一个简短的 JSP 页面来测试 Tomcat,在以后的章节里会详细讲解编写 JSP 页面的语法。

简单地说,一个 JSP 页面除了普通的 HTML 标记符外,还可以使用标记符号"<%"、"%>"加入 Java 程序片。JSP 页面文件的扩展名是 .jsp,文件的名字必须符合标识符规定,即名字可以由字母、下划线、美元符号和数字组成,并且第一个字符不能是数字字符。需要特别注意的是,文件名字中的字母是区分大小写的,如 Boy 和 boy 是不同的标识符。

为了明显地区分普通的 HTML 标记和 Java 程序片以及 JSP 标签,我们用大写字母书写普通的 HTML 标记符号。

可以用"记事本"或其他文本编辑器编辑 JSP 页面的源文件。如果使用"记事本"编辑 JSP 页面文件,在保存文件时必须将"保存类型"选择为"所有文件",将"编码"选择为"ANSI"。如果在保存文件时系统总在文件名后加上".txt"后缀,那么在保存文件时可以将文件名用""括起,如图 1.7 所示。

图 1.7 JSP 文件的保存

必须将编写好的 JSP 页面文件保存到 Tomcat 服务器的一个 Web 服务目录中。如果 Tomcat 的安装目录是 E:\Tomcat6.0,那么 Tomcat 服务器的 Web 服务目录的根目录是

E:\Tomcat6.0\webapps\Root。

为了测试 JSP 页面，可以将编写好的 JSP 页面文件保存或复制到上述根目录中。

【例 1-1】 制作一个简单的 JSP 页面 first.jsp，将其保存到根目录中。在浏览器的地址栏中输入"http://127.0.0.1:8080/first.jsp"，如果 Tomcat 服务器和 JDK 都安装正确，就会出现如图 1.8 所示的页面。

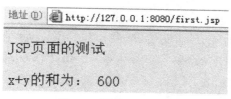

图 1.8 测试 JSP 页面

如果出现错误提示：

org.apache.jasper.JasperException: Unable to compile class for JSP

请关闭 Tomcat 服务器，然后运行 startup.bat（该文件在 Tomcat 安装目录的 bin 文件夹中），重新启动 Tomcat 服务器，以确保 Tomcat 服务器使用的 JDK 是 Java_Home 设置的 JDK。

first.jsp
```
<%@ page contentType="text/html;charset=GB2312" %>
<HTML><BODY bgcolor=cyan>
<FONT size=4>
<P>JSP 页面的测试
   <% int x=100,y=500,z;
      z=x+y;
   %>
<P>x+y 的和为：<%= z %>
</FONT></BODY>
</HTML>
```

1.4 设置 Web 服务目录

为了让客户通过浏览器访问一个 Tomcat 服务器上的 JSP 页面，必须将编写好的 JSP 页面文件保存到该 Tomcat 服务器的某个 Web 服务目录中。

1．根目录

如果 Tomcat 服务器的安装目录是 E:\Tomcat6.0，则 Tomcat 的 Web 服务目录的根目录是 E:\Tomcat6.0\webapps\Root。

用户如果准备访问根目录中的 JSP 页面，在浏览器中输入 Tomcat 服务器的 IP 地址（或域名）、端口号和 JSP 页面的名字即可（必须省略 Web 根目录的名字），如 Tomcat 服务器的 IP 地址是 192.168.1.200，根目录中存放的 JSP 页面的名字是 A.jsp，那么用户在浏览器输入的内容是"http://192.168.1.200:8080/A.jsp"。

也许用户没有为 Tomcat 服务器所在的机器设置过一个有效的 IP 地址，为了调试 JSP 页面，可以打开 Tomcat 服务器上的浏览器，在地址栏中输入"http://127.0.0.1:8080/A.jsp"。

2. 已有的 Web 服务目录

Tomcat 服务器安装目录 webapps 下的任何一个子目录都可以作为一个 Web 服务目录。

安装 Tomcat 服务器后，webapps 目录下有如下子目录：jsp-examples、balancer、servlets-examples、servlets-examples 和 tomcat-docs，也可以在目录 webapps 下再新建子目录，如子目录 Dalian。

如果将 JSP 页面文件 A.jsp 保存到目录 webapps 下的服务目录中，那么应当在浏览器的地址栏中输入 Tomcat 服务器的 IP 地址（或域名）、端口号、Web 服务目录和 JSP 页面的名字。如 A.jsp 保存到 jsp-examples 中，则输入的内容如下"http://127.0.0.1:8080/jsp-examples/A.jsp"。

3. 建立新的 Web 服务目录

可以将 Tomcat 服务器所在计算机的某个目录设置成一个 Web 服务目录，并为该 Web 服务目录指定虚拟目录，即隐藏 Web 服务目录的实际位置，用户只能通过虚拟目录访问 Web 服务目录中的 JSP 页面。

修改 Tomcat 服务器安装目录下 conf 目录中的 server.xml 文件，来设置新的 Web 服务目录。假设要将 D:\MyJsp\star 和 C:\sun 作为 Web 服务目录，并让用户分别使用 hello 和 moon 虚拟目录访问 Web 服务目录 D:\MyJsp\star 和 C:\sun 下的 JSP 页面，则应先用记事本打开文件夹 Tomcat6.0\conf 中的主配置文件 server.xml，找到出现"</HOST>"的部分（server.xml 文件尾部），然后在"</HOST>"的前面加入如下内容：

```
<Context path="/hello" docBase="D:/MyJsp/star" debug="0" reloadable="true" />
<Context path="/moon" docBase="C:/sun" debug="0" reloadable="true" />
```

注意：XML 文件是区分大小写的，不可以将<Context>写成<context>。

主配置文件 server.xml 修改后，必须重新启动 Tomcat 服务器。这样，用户就可以将 JSP 页面存放到目录 D:\MyJsp\star 或 C:\sun 中，可以通过虚拟目录 hello 或 moon 访问 JSP 页面，如将 A.jsp 保存到目录 D:\MyJsp\star 或 C:\sun 中，应在浏览器地址栏中输入"http://127.0.0.1:8080/hello/A.jsp"或"http://127.0.0.1:8080/moon/A.jsp"。

注意：在学习或使用 JSP 时，不提倡将所有的 JSP 页面都存放在 Tomcat 服务器的根目录中，应当善于建立新的 Web 应用目录，以便有效地管理 JSP 页面。

4. 相对目录

Web 服务目录下的目录称为该 Web 服务目录下的相对服务目录。例如，可以在 Web 服务目录 C:\sun 下再建立子目录 image，将文件 B.jsp 保存到 image 目录中，则可以在浏览器的地址栏中输入"http://127.0.0.1:8080/moon/image/B.jsp"来访问 B.jsp。Web 服务目录下的 JSP 页面可以通过相对路径来访问子目录中的 JSP 页面，如 Web 服务目录 sun 通过相对路径访问子目录中的 JSP 页面 image/B.jsp（不可以写成"/image/B.jsp"，"/"代表根目录，"/image/B.jsp"中的 image 将代表一个 Web 服务目录，而不再是 Web 服务目录 sun 的一个子目录）。

【例 1-2】将 A.jsp 保存到 Web 服务目录 C:\sun 中，将 B.jsp 保存到 sun 的相对目录 image 中。A.jsp 中通过超链接访问 B.jsp，如下所示：

A.jsp
```
<%@ page contentType="text/html;charset=GB2312" %>
<HTML><BODY>
```

```
          <A href=image/B.jsp > 链接到 B.jsp</A>
        </BODY></HTML>
        <%@ page contentType="text/html;charset=GB2312" %>
B.jsp
        <HTML><BODY bgcolor=cyan>
        <FONT size=8>
        <P> 1 到 100 的连续和
            <% int sum=0;
                for(int i=1;i<=100;i++){
                    sum=sum+i;
                }
                out.println("sum="+sum);
            %>
        </FONT>
        </BODY></HTML>
```

1.5 设置端口号

8080 是 Tomcat 服务器的默认端口号。可以通过修改 Tomcat 服务器的 conf 目录下的主配置文件 server.xml 来更改端口号。用记事本打开 server.xml 文件，找到如下部分：

```
<Connector port="8080"    maxHttpHeaderSize="8192"
            maxThreads="150"   minSpareThreads="25"   maxSpareThreads="75"
            enableLookups="false"   redirectPort="8443"   acceptCount="100"
            connectionTimeout="20000"   disableUploadTimeout="true"
/>
```

将其中的 port="8080" 更改为新的端口号即可，如将"8080"更改为"9080"等。

如果 Tomcat 服务器所在的计算机没有启动其他占用端口号 80 的网络程序，也可以将 Tomcat 服务器的端口号设置为 80，这样用户在访问 JSP 页面时可以省略端口号，如"http://127.0.0.1/first.jsp"。

习 题 1

1. 安装 JDK 后应当进行哪些设置？
2. 运行 startup.bat 文件启动 Tomcat 服务器的好处是什么？
3. Hello.jsp 和 hello.jsp 是否是相同的 JSP 文件名字？
4. 请在 C:\下建立一个名字为 Game 的目录，并将该目录设置成一个 Web 服务目录，然后编写一个简单的 JSP 页面，保存到该目录中，让用户使用虚拟目录 moon 访问该 JSP 页面。
5. 怎样访问 Web 服务目录子目录中的 JSP 页面？
6. 如果想修改 Tomcat 服务器的端口号，应当修改哪个文件？能否将端口号修改为 80？

第 2 章 JSP 页面

本章导读

- 知识点：掌握 JSP 页面的基本结构以及运行原理。
- 重点：学习怎样声明 JSP 页面的成员变量、方法；掌握怎样使用 Java 程序片和 Java 表达式。
- 难点：掌握 Java 程序片的执行原理。
- 关键实践：上机编写两个 JSP 页面。一个 JSP 页面重点测试 JSP 页面的成员变量和局部变量的不同，另一个 JSP 页面着重尝试怎样分解一个 Java 程序片。

2.1 JSP 页面的基本结构

在传统的 HTML 页面文件中加入 Java 程序片和 JSP 标签，就构成了一个 JSP 页面。一个 JSP 页面可由 5 种元素组合而成：① 普通的 HTML 标记符；② JSP 标记，如指令标记、动作标记；③ 成员变量和方法；④ Java 程序片；⑤ Java 表达式。

Tomcat 服务器的安装目录下 webapps 的子目录都可以作为一个 Web 服务目录，在 webapps 目录下新建一个目录 chapter2。除非特别约定，本章例子中的 JSP 页面均保存在目录 chapter2 中。

【例 2-1】 （效果如图 2.1 所示）

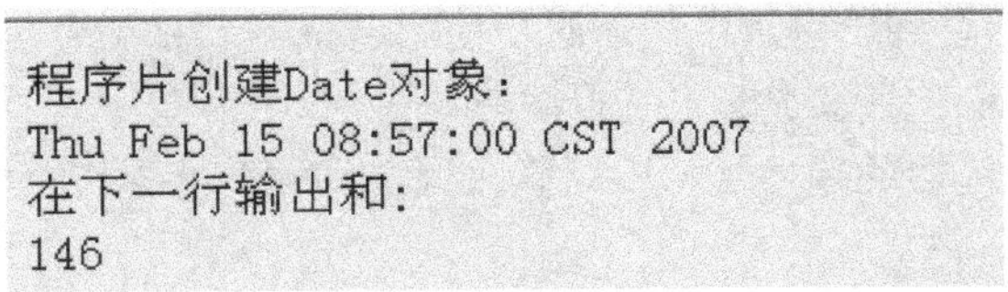

```
程序片创建Date对象：
Thu Feb 15 08:57:00 CST 2007
在下一行输出和：
146
```

图 2.1 简单的 JSP 页面

example2_1.jsp

```
<%@ page contentType="text/html;charset=GB2312" %>    <!-- JSP 指令标记 -->
<%@ page import="java.util.Date"  %>    <!-- JSP 指令标记 -->
<%!  Date date;                                         //数据声明
     int sum;
     public int add(int m,int n) {                      //方法声明
         return m+n;
     }
%>
<HTML><BODY bgcolor=cyan>    <!—HTML 标记 -->
<FONT size=4><P>程序片创建 Date 对象：
  <%  date=new Date();                                  //Java 程序片
      out.println("<BR>"+date);
      sum=add(12,34);
  %>
<BR>在下一行输出和：<BR>
   <%= sum+100 %>    <!-- Java 表达式 -->
</FONT></BODY></HTML>
```

2.2 JSP 的运行原理

当 Tomcat 服务器上的一个 JSP 页面第一次被请求执行时，Tomcat 服务器将启动一个线程。该线程的任务是，首先将 JSP 页面文件转译成一个 Java 文件，再将这个 Java 文件编译生成字节码文件，并将该字节码文件加载到内存中，然后执行字节码文件响应客户的请求，该线程完成任务后，线程进入死亡状态。这个字节码的功能如下：

- 把 JSP 页面中普通的 HTML 标记符号，交给客户的浏览器执行显示。
- JSP 标记、数据和方法声明、Java 程序片由 Tomcat 服务器负责执行，将需要显示的结果发送给客户的浏览器。
- Java 表达式由 Tomcat 服务器负责计算，将结果转化为字符串，交给客户的浏览器负责显示。

被加载到内存中的字节码将常驻内存，当这个 JSP 页面再次被请求执行时，Tomcat 服务器将再启动一个线程，直接执行常驻内存的字节码文件来响应客户。这也是 JSP 比 ASP 速度快的一个原因。而 JSP 页面的首次执行往往由服务器管理者来执行。

当多个客户请求一个 JSP 页面时，Tomcat 服务器为每个客户启动一个线程，该线程负责执行常驻内存的字节码文件来响应相应客户的请求。这些线程由 Tomcat 服务器来管理，将 CPU 的使用权在各线程之间快速切换，以保证每个线程都有机会执行字节码文件（如图 2.2 所示），这与传统的 CGI 为每个客户启动一个进程相比较，效率要高得多。

注：如果对 JSP 页面进行了修改、保存，那么 Tomcat 服务器会生成新的字节码。

下面是 Tomcat 服务器生成的 example2_1.jsp 的 Java 文件，我们把 Tomcat 服务器交给客户端浏览器负责显示的内容做了注释（§）。如果 JSP 页面保存在目录 Root 中，则在 Tomcat 服务器下的目录 work\Catalina\localhost_org\apache\jsp 中，可以找到 Tomcat 服务器生成的 JSP 页面对应的 Java 文件和编译 Java 文件得到的字节码文件。

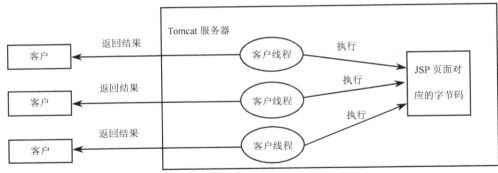

图 2.2 JSP 的运行原理

下面是 Tomcat 服务器生成的 example2_1.jsp 的 Java 文件，我们把 Tomcat 服务器交给客户端浏览器负责显示的内容做了注释（§）。如果 JSP 页面保存在目录 root 中，则在 Tomcat 服务器下的目录 work\Catalina\localhost_org\apache\jsp 中，可以找到 Tomcat 服务器生成的 JSP 页面对应的 Java 文件和编译 Java 文件得到的字节码文件。

以下是 example2_1.jsp 页面对应的 Java 文件，我们把 Tomcat 服务器返回给客户端的结果做了注释（用§标注）。

```
package org.apache.jsp;
import javax.servlet.*;
import javax.servlet.http.*;
import javax.servlet.jsp.*;
import java.util.Date;
public final class Example2_005f1_jsp extends org.apache.jasper.runtime.HttpJspBase
            implements org.apache.jasper.runtime.JspSourceDependent {
    Date date;
    int sum;
    public int add(int m,int n) {
        return m+n;
    }
    private static java.util.List _jspx_dependants;
    public Object getDependants() {
        return _jspx_dependants;
    }
    public void _jspService(HttpServletRequest request, HttpServletResponse response)
                    throws java.io.IOException, ServletException {
        JspFactory _jspxFactory = null;
        PageContext pageContext = null;
        HttpSession session = null;
        ServletContext application = null;
        ServletConfig config = null;
        JspWriter out = null;
        Object page = this;
        JspWriter _jspx_out = null;
        PageContext _jspx_page_context = null;
```

```
try {
    _jspxFactory = JspFactory.getDefaultFactory();
    response.setContentType("text/html;charset=GB2312");
    pageContext=_jspxFactory.getPageContext(this, request, response, null, true, 8192, true);
    _jspx_page_context = pageContext;
    application = pageContext.getServletContext();
    config = pageContext.getServletConfig();
    session = pageContext.getSession();
    out = pageContext.getOut();
    _jspx_out = out;
    out.write(" \r\n");                                    (§)
    out.write("\r\n");                                     (§)
    out.write("\r\n");                                     (§)
    out.write("<HTML>\r\n");                               (§)
    out.write("<BODY BGCOLOR=cyan>\r\n");                  (§)
    out.write("<FONT Size=8>\r\n");                        (§)
    out.write("<P>显示时间\r\n");                          (§)
    out.write("    ");                                     (§)
    date=new Date();
    out.println(date);
    sum=add(12,34);
    out.write("\r\n");                                     (§)
    out.write("   <BR>在下一行输出和:<BR>\r\n");           (§)
    out.write("    ");                                     (§)
    out.print( sum+100);                                   (§)
    out.write("\r\n");                                     (§)
    out.write("</FONT>\r\n");                              (§)
    out.write("</BODY>\r\n");                              (§)
    out.write("</HTML>\r\n");                              (§)
}
catch (Throwable t) {
    if (!(t instanceof SkipPageException)){
        out = _jspx_out;
        if (out != null && out.getBufferSize() != 0)
            out.clearBuffer();
        if (_jspx_page_context != null)
            _jspx_page_context.handlePageException(t);
    }
}
finally {
    if (_jspxFactory != null)
        _jspxFactory.releasePageContext(_jspx_page_context);
}
}
}
```

2.3 JSP 页面的成员变量和方法

我们已经知道，JSP 页面可包含 HTML 标记、JSP 指令标记、成员变量和方法、Java 程序片和 Java 表达式，本节介绍 JSP 页面中的成员变量和方法。

JSP 页面在标记符"<%!"和"%>"之间声明它的成员变量和方法。

1．声明变量

可以在 JSP 页面的标记符"<%!"和"%>"之间声明变量，即在"<%!"和"%>"之间放置 Java 的变量声明语句，变量的类型可以是 Java 语言允许的任何数据类型。

标记符"<%!"和"%>"之间被声明的变量称为 JSP 页面的成员变量。例如：

```
<%!  int x, y=100;
     Date date;
%>
```

"<%!"和"%>"之间声明的变量在整个 JSP 页面内有效，因为 Tomcat 服务器将 JSP 页面转译成 Java 文件时，将这些变量作为类的成员变量，这些变量的内存空间直到服务器关闭才释放。当多个客户请求一个 JSP 页面时，Tomcat 服务器为每个客户启动一个线程，这些用户线程将共享 JSP 页面的成员变量。这些用户线程由 Tomcat 服务器来管理，Tomcat 服务器将 CPU 的使用权在各线程间快速切换，以保证每个线程轮流执行 JSP 页面（对应的字节码），因此任何一个线程对 JSP 页面成员变量操作的结果，都会影响到其他线程。

【例 2-2】 利用成员变量被所有用户共享这一性质，实现一个简单的计数器（效果如图 2.3 所示）。

example2_2.jsp
```
<%@ page contentType="text/html;charset=GB2312" %>
<HTML><BODY bgcolor=pink><FONT size=4>
  <%!  int i=0;
  %>
  <%   i++;
  %>
<P>您是第<%=i%>个访问本站的客户。
</BODY></HTML>
```

您是第58个访问本站的客户。

图 2.3　成员变量的使用

2．声明方法

JSP 页面在"<%!"和"%>"之间声明定义若干个方法，这些方法可以在 Java 程序片中被调用执行。在例 2-3 中，我们在"<%!"和"%>"之间声明定义了两个方法：add(int x,int y) 和 sub(int x,int y)，然后在程序片中调用这两个方法。

【例 2-3】 （效果如图 2.4 所示）

example2_3.jsp

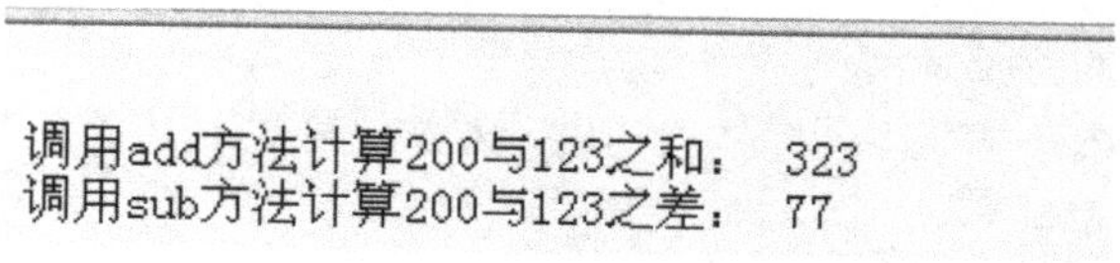

调用add方法计算200与123之和： 323
调用sub方法计算200与123之差： 77

图 2.4 声明与使用方法

```
<%@ page contentType="text/html;Charset=GB2312" %>
<HTML><BODY bgcolor=yellow>
  <%! int add(int x,int y) {
        return x+y;
      }
      int sub(int x,int y) {
        return x-y;
      }
  %>
  <%  out.println("<BR>调用 add 方法计算 200 与 123 之和：");
      int a=add(200,123);
      out.println(a);
      out.println("<BR>调用 sub 方法计算 200 与 123 之差：");
      int b=sub(200,123);
      out.println(b);
  %>
</BODY></HTML>
```

2.4 JSP 页面中的 Java 程序片

可以在"<%"和"%>"之间插入 Java 程序片。一个 JSP 页面可以有许多程序片，这些程序片按顺序执行。在一个程序片中声明的变量称为 JSP 页面的局部变量。该局部变量在 JSP 页面后继的所有程序片以及表达式部分内都有效。这是因为，Tomcat 服务器将 JSP 页面转译成 Java 文件时，将程序片中声明的变量作为类中某个方法的变量（即局部变量）使用。

当多个客户请求一个 JSP 页面时，Tomcat 为每个客户启动一个线程，然后该线程执行 JSP 页面，那么 Java 程序片将被执行多次，分别在不同的线程中执行，如图 2.5 所示。

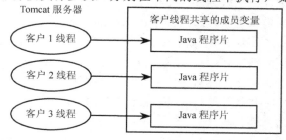

图 2.5 程序片的执行

Java 程序片的执行具有如下特点。

（1）操作 JSP 页面的成员变量

Java 程序片中操作的成员变量是各线程共享的变量，任何一个线程对 JSP 页面成员变量操作的结果，都会影响到其他线程。

（2）调用 JSP 页面的方法

Java 程序片中可以出现方法调用语句，该方法必须是 JSP 页面的方法。

（3）声明操作局部变量

当一个线程享用 CPU 资源时，Tomcat 让该线程执行程序片。这时，程序片中的局部变量被分配内存空间，当轮到另一个线程享用 CPU 资源时，Tomcat 让该线程再次执行 Java 程序片，那么，程序片中的局部变量会再次分配内存空间。也就是说，程序片已经被执行了两次，分别运行在不同的线程中，即运行在不同的时间片内。运行在不同线程中的程序片的局部变量互不干扰，即一个用户改变 Java 程序片中的局部变量的值不会影响其他用户的程序片中的局部变量。

当一个线程的程序片执行完毕，运行在该线程中的程序片的局部变量释放所占的内存。

【例 2-4】 计算 1～100 的连续和。

example2_4.jsp

```
<%@ page contentType="text/html;charset=GB2312" %>
<HTML><BODY bgcolor=cyan><FONT size=6>
    <%! int continueSum(int n) {
            int sum=0;
            for(int i=1;i<=n;i++){
                sum=sum+i;
            }
            return sum;
        }
    %>
    <P> 1～100 的连续和：<BR>
    <%  long sum;
        sum=continueSum(100);
        out.print(sum);
    %>
</BODY></HTML>
```

当多个客户同时请求一个 JSP 页面时，JSP 页面的程序片就会被多次调用运行，分别运行在不同的线程中。注意，一个线程在执行程序片期间可以调用 JSP 页面的方法操作 JSP 页面的成员变量，而这些成员变量是被所有的用户共享的。在编写 JSP 页面时，对这种情况应多加注意。例如，如果一个用户在执行程序片时调用 JSP 页面的方法操作成员变量时，可能不希望其他用户也调用该方法操作成员变量，以免对其产生不利的影响，就应该将操作成员变量的方法用 synchronized 关键字修饰。当一个线程在执行程序片期间调用 synchronized 方法时，其他线程想在程序片中调用该 synchronized 方法时就必须等待，直到调用 synchronized 方法的线程调用执行完该方法。

【例 2-5】 通过 synchronized 方法操作一个成员变量来实现一个简单的计数器。

example2_5.jsp

```
<%@ page contentType="text/html;Charset=GB2312" %>
<HTML>
<BODY>
    <%! int count=0;                        //被客户共享的 count
        synchronized void setCount(){       //synchronized 修饰的方法
```

```
                    count++;
                }
        %>
        <%   setCount();
                out.println("您是第"+count+"个访问本站的客户");
        %>
</BODY></HTML>
```

有时根据需要,可以将程序片分成几个小的程序片,以便插入 JSP 标记或 HTML 标记。

【例 2-6】 在几个程序片之间插入 HTML 标记(效果如图 2.6 所示)。

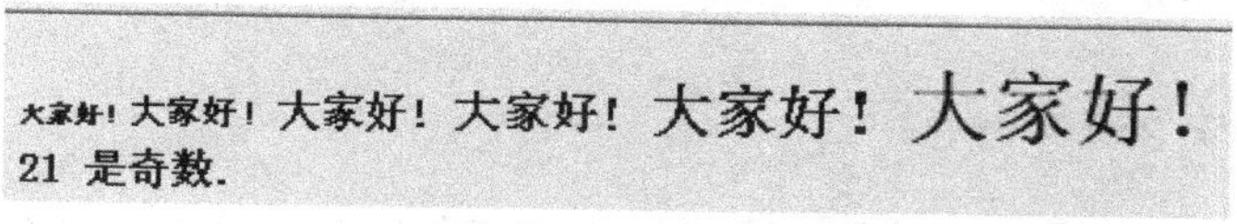

图 2.6　程序片的分解

example2_6.jsp
```
        <%@ page contentType="text/html;Charset=GB2312" %>
        <HTML><BODY bgcolor=cyan>
    <%   int sum=0,i,j;
            for(i=1;i<=6;i++){
    %>        <Font size=<%=i%>>
                <b>大家好!</b>
                </Font>
    <%       sum=sum+i;
            }
            if(sum%2==0) {
                out.print("<BR><b>"+sum+"</b>");
    %>        <b>是偶数.</b>
    <%      }
            else{
                out.print("<BR><b>"+sum+"</b>");
    %>        <b>是奇数.</b>
    <%      }
    %>
</BODY></HTML>
```

2.5　JSP 页面中的 Java 表达式

"<%=" 和 "%>" 之间可以是一个 Java 表达式(注意:"<%=" 是一个完整的符号,"<%" 和 "=" 之间不要有空格),这个表达式必须能求值。表达式的值由服务器负责计算,并将计算结果用字符串形式发送到客户端显示。

【例 2-7】 计算表达式的值。

example2_7.jsp
```
        <%@ page contentType="text/html;charset=GB2312" %>
        <HTML><BODY  bgcolor=cyan><FONT size=4>
        <P> Sin(0.9)除以 3 等于
```

```
            <%=Math.sin(0.90)/3%>
     <p>3 的平方是:
            <%=Math.pow(3,2)%>
     <P>12345679 乘 72 等于
            <%=12345679*72%>
     <P> 5 的平方根等于
            <%=Math.sqrt(5)%>
     <P>99 大于 100 吗？回答：
            <%=99>100%>
     </BODY></HTML>
```

2.6 JSP 中的注释

注释可以增强 JSP 文件的可读性，便于 JSP 文件的维护。常见的 JSP 中的注释有两种。

（1）HTML 注释

在标记符号 "<!--" 和 "-->" 之间加入注释内容：

    <!--   注释内容   -->

JSP 引擎把 HTML 注释交给客户机，客户机通过浏览器查看 JSP 的源文件时，能够看到 HTML 注释。

（2）JSP 注释

在标记符号 "<%--" 和 "--%>" 之间加入注释内容：

    <%--   注释内容   --%>

JSP 引擎忽略 JSP 注释，即在编译 JSP 页面时忽略 JSP 注释。客户通过浏览器查看 JSP 的源文件时，不能看到 JSP 注释。

【例 2-8】 使用注释。

example2_8.jsp
```
     <%@ page contentType="text/html;charset=GB2312" %>
     <HTML><BODY>
        <!-- 以下是标题 1 -->
        <H1> 大家要注意了，成员变量是大家共享的 </H1>
        <%-- 下面是变量和方法的声明，利用共享的 sum 和 n 计算连续和 --%>
        <%! long sum=0,n=1;
            void computer(){
               sum=sum+n;
               n=n+1;
            }
        %>
        <%-- 下面是程序片，调用方法得到结果 --%>
        <%  computer();
            out.println("当前连续和是："+sum);
        %>
     </BODY></HTML>
```

2.7 在 JSP 页面中使用 HTML 标记

JSP 页面可以含有 HTML 标记，当客户通过浏览器请求一个 JSP 页面时，Tomcat 服务器将该 JSP 页面中的 HTML 标记直接发送给客户机的浏览器，由客户机的浏览器负责执行这些 HTML 标记。而 JSP 页面中的变量声明、程序片以及表达式由 Tomcat 服务器处理后，再将有关的结果用文本方式发送给客户机的浏览器。

JSP 页面中的 HTML 标记是页面的静态部分，即不需要服务器做任何处理，直接发送给客户机的信息。通过使用 HTML 标记，JSP 页面可以为用户提供一个友好的界面，即数据表示层。而 JSP 页面中的变量声明、程序片和表达式为动态部分，需要服务器做出处理后，再将有关处理后的结果发送给客户。

编写一个健壮的 Web 应用程序，提倡将数据的表示和处理分离。如果将数据表示和处理混杂在一个 JSP 页面中，将导致代码混乱，不利于 Web 应用的拓展和维护。学习 JSP，有必要了解和掌握在 JSP 页面中怎样使用 Java 程序片来处理数据，因为 JSP 页面在必要时可以使用少量的 Java 程序片，而且早期的许多 Web 应用程序中的 JSP 页面中有大量的 Java 程序片，这些项目可能需要维护或修改。在后续的章节中，我们将学习怎样使用 JavaBean 和 Servlet 来分离数据的表示和处理。

HTML 是 HyperText Marked Language 的缩写，即超文本标记语言。用 HTML 编写的文件扩展名为 .html（或.htm），称为超文本文档。JSP 页面中可以使用 HTML 标记来显示数据，如 "
你好</br>" 将在一个新行中显示 "你好"，"<H1>你好</H1>" 将用黑体一号字显示 "你好"。目前的 HTML 大约有 100 多个标记，这些标记可以描述数据的显示格式，如果读者对 HTML 比较陌生，建议补充这方面的知识。本节将简单介绍一些重要的 HTML 标记。

1. 表单标记

由于客户机经常需要使用 HTML 标记中的表单标记来提交数据，所以有必要对表单标记进行简明介绍。

表单的一般格式如下：

```
<FORM   method= get| post   action="提交信息的目的地页面"   name="表单的名字">
      数据提交手段部分
</FORM>
```

其中，<FORM>是表单标记，method 取值 get 或 post。get 方法与 post 方法的主要区别是：get 方法提交的信息会在提交的过程中显示在浏览器的地址栏中，而 post 方法提交的信息不会显示在地址栏中。提交手段包括：通过文本框、通过列表、通过文本区等。例如：

```
<FORM   action="tom.jsp" method= "post" >
      <Input   type="text" name="game" value= "ok" >
      <Input   type ="submit" value="送出" name="submit" >
</FORM>
```

一个表单的数据提交手段部分经常包括如下子标记符号：

- <Input … >
- <Select … >   </Select>
- <Option … >   </Option>

⊙ <TextArea ... >   </TextArea>

2．<Input>标记

在表单中，用 Input 标记来指定表单中数据的输入方式以及表单的提交键。Input 标记中的 type 属性可以指定输入方式的 GUI 对象，name 属性用来指定这个 GUI 对象的名称。其基本格式如下：

    <Input type="输入对象的 GUI 类型" name= "名字" >

服务器通过属性 name 指定的名字来获取"输入对象的 GUI 类型"中提交的数据。"输入对象的 GUI 类型"可以是 text（文本框）、checkbox（复选框）、submit（提交键）、hidden（隐藏）等。

（1）text

当输入对象的 GUI 类型是 text 时，除了用 name 为 text 指定名字外，还可以为 text 指定其他的一些值。例如：

    <Input type="text" name="me" value="hi" size="12" algin="left" maxlength="30">

其中，value 的值是 text 的初始值，size 是 text 对象的长度（单位是字符），algin 是 text 在浏览器窗体中的对齐方式，maxlength 指定 text 可输入字符的最大长度。

（2）radio

当输入对象的 GUI 类型是 radio 时，除了用 name 为 radio 指定名字外，还可以为 radio 指定其他一些值。例如：

    <Input type="radio" name="rad" value="red" algin="top" checked="Java">

其中，value 指定 radio 的值；algin 是 radio 在浏览器窗体中的对齐方式；如果几个单选钮的 name 取值相同，那么同一时刻只能有一个被选中；服务器通过 name 指定的名字来获取被选中的 radio 提交的由 value 指定的值；checked 如果取值是一个非空的字符串，那么该单选框的初始状态就是选中状态。

（3）checkbox

当输入对象的 GUI 类型是 checkbox 时，除了用 name 为 checkbox 指定名字外，还可以为 checkbox 指定其他值。例如：

    <Input type="checkbox" name="ch" value="pink" algin="top" checked="Java">

其中，value 指定 checkbox 的值。复选框与单选钮的区别就是复选框可以多选。服务器通过 name 指定的名字来获取被选中的 checkbox 提交的由 value 指定的值。checked 如果取值是一个非空的字符串，那么该复选框的初始状态就是选中状态。

（4）password

password 是输入口令用的特殊文本框，输入的信息用"*"回显，防止他人偷看。例如：

    <Input type="password" name="me" size="12" maxlength="30">

服务器通过 name 指定的字符串获取 password 提交的值，在口令框中输入"12345"，那么"12345"将被提交给服务器。口令框仅仅起着不让别人偷看的作用，不提供加密保护措施。

（5）hidden

当<Input>中的属性 type 的值为 hidden 时，<Input>没有可见的输入界面，表单直接将<Input>中 value 属性的值提交给服务器。例如：

    <Input type="hidden" name="h" value="123">

服务器通过 name 指定的名字来获取由 value 指定的值。

（6）submit

为了能把表单的数据提交给服务器，一个表单至少要包含一个提交键。例如：

    \<Input type="submit" name="me" value="ok" size="12"\>

单击【提交】按钮后，服务器就可以获取表单提交的各数据。当然，服务器也可以获取提交键的值，服务器通过 name 指定的名字来获取提交键提交的由 value 指定的值。

（7）reset

重置键将表单中输入的数据清空，以便重新输入数据。例如：

    \<Input type="reset"\>

【例 2-9】 JSP 页面 input.jsp 用表单向 receive.jsp 页面提交数据，input.jsp 和 receive.jsp 均保存在 Web 服务目录 chaper2 中（效果如图 2.7 和图 2.8 所示）。

图 2.7　使用表单提交数据 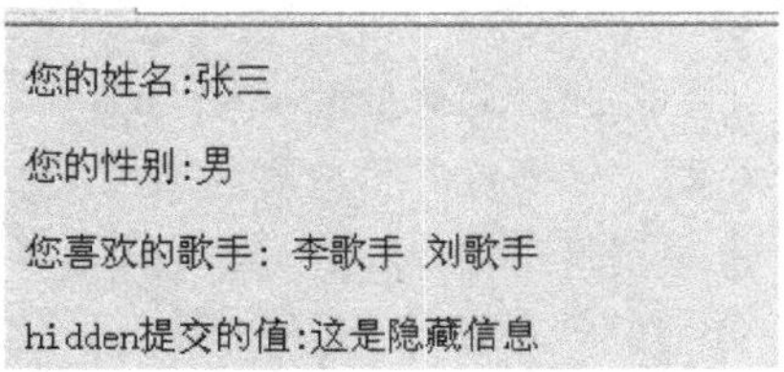　　　　　　图 2.8　获取表单提交的数据

input.jsp

```
<%@ page contentType="text/html;Charset=GB2312" %>
<HTML><BODY bgcolor=cyan><FONT size=3>
    <FORM action="receive.jsp" method=post name=form>
    <P>请输入下列信息：
        <BR>输入您的姓名： <Input type="text" name="name" value="张三"></BR>
        <BR>选择性别： <Input type="radio" name="R" value="男" checked="default">男
                      <Input type="radio" name="R" value="女">女
        </BR>
        <BR>选择您喜欢的歌手:
           <Input type="checkbox" name="superstar" value="张歌手" >张歌手
           <Input type="checkbox" name="superstar" value="李歌手" >李歌手
           <Input type="checkbox" name="superstar" value="刘歌手" >刘歌手
           <Input type="checkbox" name="superstar" value="王歌手" >王歌手
        </BR>
        <Input type="hidden" value="这是隐藏信息" name="secret">
        <Input type="submit" value="提交" name="submit">
    </FORM>
</FONT></BODY></HTML>
```

receive.jsp

```
<%@ page contentType="text/html;Charset=GB2312" %>
<HTML><BODY bgcolor=cyan><FONT size=3>
    <%  String yourName=request.getParameter("name");        //获取 text 提交的值
        String yourSex=request.getParameter("R");            //获取 radio 提交的值
        String secretMess=request.getParameter("secret");    //获取 hidden 提交的值
```

```
            //获取checkbox提交的值
            String personName[]=request.getParameterValues("superstar");
            out.println("<P> 您的姓名:"+yourName+"</P>");
            out.println("<P> 您的性别:"+yourSex+"</P>");
            out.println("<P> 您喜欢的歌手:");
            if(personName==null) {
                out.print("一个都不喜欢");
            }
            else{
                for(int k=0;k<personName.length;k++){
                    out.println(" "+personName[k]);
                }
            }
            out.println("<P> hidden提交的值:"+secretMess);
        %>
    </FONT></BODY></HTML>
```

注意： 上面的例子中要特别注意 "<%@ page contentType="text/html;Charset= GB2312"%>" 中出现的 "Charset" 的大小写拼写，不要写成 "charset"，否则容易出现中文乱码。中文乱码问题将在本书 4.1 节中介绍。

3．<Select>、<Option>标记

在表单中，可以使用下拉列表或滚动列表来选择要提交的数据。下拉式列表和滚动列表通过在<Select>标记中使用若干个<Option>子标记来定义。其基本格式如下：

```
<Select   name="下拉列表的名字">
    <Option   value="cat">文本信息
    <Option   value="dog">文本信息
    ……
    <Option   value="600">文本信息
</Select>
```

服务器通过属性 name 的值获取下拉列表中被选中的 Option 的值（参数 value 指定的值）。

在 Select 标记中增加 size 属性，就变成了滚动列表。size 的值确定滚动列表中选项的可见数目。滚动列表的基本格式如下：

```
<Select   name="下拉列表的名字" size="一个正整数">
    <Option   value="cat">文本信息
    <Option   value="dog">文本信息
    ……
    <Option   value="600">文本信息
</Select>
```

服务器通过属性 name 的值获取滚动列表中被选中的 option 的值（参数 value 指定的值）。

4．<TextArea>标记

<TextArea>是一个能输入或显示多行文本的文本区。在表单中，使用<TextArea>作为子标记，能使用户提交多行文本给服务器。其基本格式如下：

```
<TextArea   name="名字"   Rows="文本可见行数"   Cols="文本可见列数">
</TextArea>
```

5. 表格标记<TABLE>

表格由<TABLE>、</TABLE>标记定义，一般格式如下：
```
<TABLE>
    <TR   width="该行的宽度">
        <TH   width="单元格的宽度">单元格中的数据</TH>
        ……
        <TD   width="单元格的宽度">单元格中的数据</TD>
    </TR>
    ……
</TABLE>
```
其中，<TR>…</TR>定义表格的行，行标记通过使用<TH>或<TD>子标记来定义该行的单元格，<TH>标记定义的单元格中的数据加粗显示，<TD>不加粗显示。在<TABLE>中增加border属性，可指定该表格所带边框的宽度，如<TABLE border=2>。

【例2-10】 用一个3行的表格显示数据（效果如图2.9所示）。

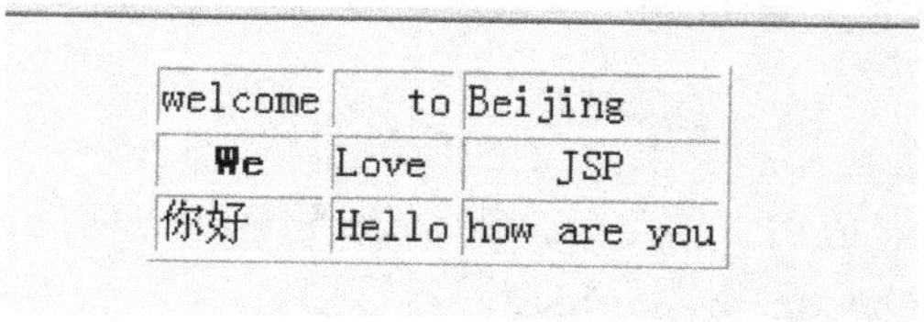

图2.9 使用表格

example2_10.jsp
```
<%@ page contentType="text/html;charset=GB2312" %>
<HTML><BODY bgcolor=yellow>
    <TABLE align="Center" border=1>
        <TR width=400>
            <TD   align="Center">welcome</TD>
            <TD   align="Right">to</TD>
            <TD   align="Left">Beijing</TD>
        </TR>
        <TR >
            <TH   valign="Top">We</TH>
            <TD   valign="Bottom">Love</TD>
            <TD   valign="Bottom" align="Center">JSP</TD>
        </TR>
        <TR>
            <TD   valign="Top">你好</TD>
            <TD   valign="Bottom">Hello</TD>
            <TD   valign="Bottom" align="Center">how are you</TD>
        </TR>
    </TABLE>
</BODY></HTML>
```

6. 图像标记<IMG>

使用图像标记可以显示一幅图像，<IMG>标记的基本格式如下：
```
<IMG   src="图像文件的 URL">描述文字</IMG>
```

如果图像文件和当前页面在同一 Web 服务目录中，图像文件的地址就是该图像文件的名字；如果图像文件在当前 Web 服务目录的一个子目录中，如 image 子目录中，那么"图像文件的 URL"就是"image/图像文件的名字"。

<IMG>标记中可以使用 width 和 height 属性指定被显示图像的宽和高。如果省略 width 和 height 属性，<IMG>标记将按图像的原始宽度和高度来显示图像。

7．多媒体标记<EMBED>

使用<EMBED>标记可以播放音乐和视频，当浏览器执行到该标记时，会把浏览器所在机器中的默认播放器嵌入到浏览器中，以便播放音乐或视频文件。其基本格式如下：

    <EMBED　src="音乐或视频文件的 URL">描述文字</EMBED>

如果音乐或视频文件和当前页面在同一 Web 应用目录中，音乐或视频文件的地址就是该文件的名字。

<EMBED>标记中经常使用的属性及取值如下：

- autostart ——取值 true 或 false，用来指定音乐或视频文件传送完毕后是否立刻播放，默认值是 false。
- loop ——取值为正整数，用来指定音乐或视频文件重复播放的次数。
- width, heigh ——取值均为正整数，用 width 和 height 属性的值指定播放器的宽和高，如果省略 width 和 height 属性，将使用默认值。

【例 2-11】 使用标记<IMG>和<EMBED>。用户通过 select.jsp 页面中的下拉列表选择一幅图像的名字，通过滚动列表选择一个视频文件或音乐文件，单击【提交】按钮，将数据提交给 show.jsp 页面，该页面使用<IMG>标记显示图像、使用<EMBED>标记播放音乐和视频（效果如图 2.10 和图 2.11 所示）。

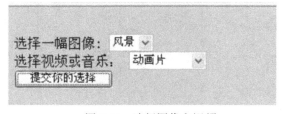 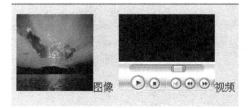

    图 2.10　选择图像和视频   图 2.11　显示图像、播放视频

select.jsp

```
<%@ page contentType="text/html;charset=GB2312" %>
<HTML><BODY bgcolor=cyan><FONT size=3>
  <FORM action="show.jsp" method=post name=form>
    <BR>选择一幅图像：
      <Select name="image" >
          <Option Selected value="img1.JPG">人物
          <Option value="img2.jpg">风景
          <Option value="img3.jpg">动物
      </Select>
    <BR>选择视频或音乐：
      <Select name="video" >
```

```
                <Option value="a.wmv">茉莉花
                <Option value="b.wmv">动画片
                <Option value="c.avi">飞翔的鸽子
            </Select>
        <BR> <Input TYPE="submit" value="提交你的选择" name="submit">
        </FORM>
    </FONT></BODY></HTML>
```

show.jsp
```
    <%@ page contentType="text/html;charset=GB2312" %>
    <HTML><BODY bgcolor=cyan><FONT size=3>
    <% String s1=request.getParameter("image");
       String s2=request.getParameter("video");
       if(s1==null) {s1="img1.JPG";}
       if(s2==null) {s2="a.avi";}
    %>
    <IMG src=<%=s1 %> width=120 height=120 >图像</IMG>
    <EMBED src=<%=s2 %> width=150 height=120 >视频</EMBED>
    </FONT></BODY></HTML>
```

习 题 2

1. "<%!"和"%>"之间声明的变量与"<%"和"%>"之间声明的变量有何不同？
2. 如果有两个用户访问一个 JSP 页面，该页面中的 Java 程序片将被执行几次？
3. 假设有两个用户访问下列 JSP 页面 test.jsp，第一个访问和第二个访问 test.jsp 页面的用户看到的页面的效果有何不同？

   test.jsp
   ```
       <%@ page contentType="text/html;charset=GB2312" %>
       <HTML><BODY>
       <%!  String str=new String("你好");
            synchronized void f(String s) {
                str=str+s;
            }
       %>
       <%  String s="abcd";
           f(s);
       %>
       <%=str%>
       </BODY></HTML>
   ```

4. 请编写一个简单的 JSP 页面，计算出 100 以内的素数。
5. 请编写两个 JSP 页面 a.jsp 和 b.jsp。a.jsp 页面使用表单提交数据给 b.jsp 页面，要求 a.jsp 通过 text 方式提交一个字符串给 b.jsp，b.jsp 页面获取 a.jsp 提交的字符串，并使用 Java 表达式显示这个字符串及其长度（所含字符的个数）。
6. 请编写 JSP 页面 inputNumber.jsp 和 getNumer.jsp。inputNumber.jsp 页面使用表单提交数据给 getNumber.jsp 页面，要求 inputNumber.jsp 通过 text 方式提交一个数字给 getNumer.jsp。

getNumer.jsp 计算并显示这个数的平方和立方。getNumer.jsp 中 Java 程序片的部分代码如下：

```
String s=request.getParameter("number");
try{
    double d=Double.parseDouble(s);
    out.println(d*d);
    out.println("<BR>"+d*d*d);
}
catch(NumberFormatException exp) {
    out.println("<BR>"+exp);
}
```

第 3 章 JSP 标记

> **本章导读**
> ✿ 知识点：掌握 JSP 指令标记、动作标记和自定义标记。
> ✿ 重点：page 指令标记、include 指令标记，include 动作标记以及 Tag 标记。
> ✿ 难点：Tag 文件的编写、保存以及 Tag 标记的使用。
> ✿ 关键实践：编写 JSP 页面，使用 Tag 标记实现代码的复用。

JSP 标记是 JSP 页面中很重要的组成部分，JSP 标记包括指令标记、动作标记和自定义标记。本章主要讲述指令标记、动作标记和自定义标记，其中自定义标记主要讲述与 Tag 文件有关的 Tag 标记。

我们在 Tomcat 服务器的目录 webapps 下新建一个 Web 服务目录 chapter3。除非特别约定，本章例子中涉及的 JSP 页面均保存到 chapter3 目录中。

3.1 指令标记 page

page 指令标记，简称 page 指令，用来定义整个 JSP 页面的一些属性和这些属性的值。可以用一个 page 指令指定多个属性的值，也可以使用多个 page 指令分别为每个属性指定值。page 指令的作用对整个页面有效，与其书写的位置无关，习惯把 page 指令写在 JSP 页面的最前面。page 指令的格式如下：

<%@ page 属性1="属性1的值" 属性2="属性2的值" ... %>

page 指令标记可以为下列属性指定值：contentType 属性，import 属性，language 属性，session 属性，buffer 属性，auotFlush 属性，isThreadSafe 属性，pageEncoding 属性，下面将分别讲述上述属性值的设置与作用。

1. 属性 contentType 的值

JSP 页面使用 page 指令只能为 contentType 指定一个值，以此确定响应的 MIME 类型。当用户请求一个 JSP 页面时，Tomcat 服务器负责解释执行 JSP 页面，并将某些信息发送给客户机的浏览器，以便用户浏览这些信息。如果希望客户机的浏览器用 HTML 解析器来解析执行这些信息，就可以用 page 指令指定 JSP 页面的 contentType 属性的值为 "text/html;charset=GB2312"，例如：

```
<%@ page contentType="text/html;charset=GB2312" %>
```

如果不使用 page 指令为 contentType 指定一个值，那么属性 contentType 的默认值是 "text/html ; charset=ISO-8859-1"。

注：不允许两次使用 page 指令给属性 contentType 指定不同的属性值，下列用法错误：

```
<%@ page contentType="text/html;charset=GB2312" %>
<%@ page contentType="application/msword" %>
```

注：如果想深入了解 MIME 类型，可以在网络搜索引擎中搜索 MIME 关键字。

【**例 3-1**】制作 3 个 JSP 页面：A.jsp 页面，使用 page 指令设置 contentType 属性的值为 "text/html;charset=GB2312"，当客户请求 A.jsp 页面时，客户的浏览器启用 HTML 解析器来解析执行收到的信息；B.jsp 页面，使用 page 指令设置 contentType 属性的值为 "application/msword"，当客户请求 B.jsp 页面时，客户的浏览器将启动本地的 Word 应用程序来解析执行收到的信息；C.jsp 页面，使用 page 指令设置 contentType 属性的值为 "image/jpeg"，当客户请求 C.jsp 页面时，客户的浏览器将启动图形解码器来解析执行收到的信息（效果如图 3.1～图 3.3 所示）。

图 3.1　text/html　　　　图 3.2　application/msword　　　　图 3.3　image/jpeg

A.jsp

```
<%@ page contentType="text/html;Charset=GB2312" %>
<HTML><BODY bgcolor=cyan>
<FONT size=3>
<P>我在学习 page 指令
</FONT></BODY></HTML>
```

B.jsp

```
<%@ page contentType="application/msword" %>
<HTML><BODY bgcolor=cyan>
<FONT size=3>
<P>启动 Word 应用程序
</FONT></BODY></HTML>
```

C.jsp

```
<%@ page contentType="image/jpeg" %>
<%@ page import="java.awt.*" %>
```

```
<%@ page import="java.io.*" %>
<%@ page import="java.awt.image.*" %>
<%@ page import="java.awt.geom.*" %>
<%@ page import="com.sun.image.codec.jpeg.*" %>
<%   int width=260, height=260;
    BufferedImage image = new BufferedImage(width,height,BufferedImage.TYPE_INT_RGB);
    Graphics g = image.getGraphics();
    g.setColor(Color.white);
    g.fillRect(0, 0, width, height);
    Graphics2D g_2d=(Graphics2D)g;
    Ellipse2D ellipse=new Ellipse2D. Double (40,80,100,40);
    g_2d.setColor(Color.blue);
    AffineTransform trans=new   AffineTransform();
    for(int i=1;i<=12;i++){
       trans.rotate(30.0*Math.PI/180,90,100);
       g_2d.setTransform(trans);
       g_2d.draw(ellipse);
    }
    g.dispose();
    OutputStream outClient= response.getOutputStream();         //获取指向客户端的输出流
    JPEGImageEncoder encoder=JPEGCodec.createJPEGEncoder(outClient);
    encoder.encode(image);
%>
```

2．属性 import 的值

import 属性的作用是为 JSP 页面引入 Java 核心包中的类，这样就可以在 JSP 页面的程序片部分、变量及方法声明部分、表达式部分使用包中的类。可以为该属性指定多个值，该属性的值可以是 Java 某包中的所有类或一个具体的类。

使用 page 指令可以为 import 属性指定几个值，这些值用逗号分隔。例如：

    `<%@ page   import="java.util.*" ,"java.io.*" , "java.awt.*" %>`

也可以使用多个 page 指令为 import 属性指定几个值。例如：

    `<%@ page import=" java.util..*" %>`
    `<%@ page import="java.io.*" %>`
    `<%@ page import="java.awt.*" %>`

当为 import 指定多个属性值时，JSP 引擎把 JSP 页面转译成的 Java 文件中会有如下 import 语句：

    import java.util.*;
    import java.io.*;
    import java.awt.*;

JSP 页面的默认 import 属性有如下值："java.lang.*"、"javax.servlet.*"、"javax.servlet.jsp.*"、"javax.servlet.http.*"。import 属性是一个比较特殊的属性，可以被指定多个值。需要特别注意的是，除 import 属性外，其余属性只能被指定一个值。

3．属性 language 的值

language 属性定义 JSP 页面使用的脚本语言，该属性的值目前只能取"java"。为 language 属性指定值的格式如下：

    `<%@ page   language="java" %>`

language 属性的默认值是"java"。如果在 JSP 页面中没有使用 page 指令指定该属性的值，那么 JSP 页面默认有如下 page 指令：

  <%@ page   language="java" %>

4．属性 session

  session 属性的值用于设置是否需要使用 Tomcat 服务器内置的 session 对象。session 属性的属性值可以是 true 或 false，默认属性值是 true。

5．属性 buffer

  内置输出流对象 out 负责将服务器的某些信息或运行结果发送到客户端显示，buffer 属性用来指定 out 使用的缓冲区大小或不使用缓冲区。

  buffer 属性的默认值是 8KB，可以使用 page 指令修改这个默认值。例如：

  <%@ page buffer= "24KB" %>

  如果不准备让 out 流使用缓冲区，可将 buffer 属性的值设置为 none。

6．属性 autoFlush

  autoFlush 属性指定 out 流的缓冲区被填满时，缓冲区是否自动刷新。

  autoFlush 可以取值 true 或 false，默认值是 true。当 autoFlush 属性取值为 false 时，如果 out 的缓冲区填满，就会出现缓存溢出异常。当 buffer 的值为 none 时，autoFlush 的值就不能设置成 false。

7．属性 isThreadSafe

  isThreadSafe 属性取值 true 或 false，默认值是 true。Tomcat 服务器使用多线程技术处理客户的请求，即当多个客户机请求一个 JSP 页面时，Tomcat 服务器为每个客户机启动一个线程，每个线程分别负责执行常驻内存的字节码文件来响应相应客户机的请求。这些线程由 Tomcat 服务器来管理，将 CPU 的使用权在各线程之间快速切换，以保证每个线程都有机会执行字节码文件。当 isThreadSafe 属性值为 true 时，CPU 的使用权在各线程间快速切换，也就是说，即使一个客户机的线程没有执行完毕，CPU 的使用权也可能要切换给其他线程。如此轮流，直到各线程执行完毕。当 JSP 使用 page 指令将 isThreadSafe 属性值设置成 false 时，该 JSP 页面同一时刻只能处理响应一个客户机的请求，其他客户机需排队等待。也就是说，CPU 要保证一个线程将 JSP 页面执行完毕，才会把 CPU 使用权切换给其他线程。

3.2 指令标记 include

  设计一个 Web 应用可能需要编写若干个 JSP 页面，如果这些 JSP 页面都需要显示某些同样的信息，如每个 JSP 页面上都可能需要一个导航条，以便用户在各 JSP 页面之间切换，那么每个 JSP 页面都可以使用 include 指令标记在页面的适当位置整体嵌入一个相同的文件。

  指令标记 include 的语法如下：

  <%@ include file= "文件的 URL" %>

  该指令标记的作用是在 JSP 页面上出现该指令的位置处静态嵌入一个文件。被嵌入的文件必须是可访问和可使用的。如果该文件和当前 JSP 页面在同一 Web 服务目录中，那么"文件的 URL"就是文件的名字；如果该文件在 JSP 页面所在的 Web 服务目录的一个子目录中，

如 image 子目录，那么"文件的 URL"就是"image/文件的名字"。

所谓静态嵌入，就是 Tomcat 服务器在编译阶段就完成文件的嵌入操作，即将当前的 JSP 页面和要嵌入的文件合并成一个新的 JSP 页面，Tomcat 服务器再将这个新的 JSP 页面转译成 Java 文件。因此，嵌入文件后，必须保证新合并成的 JSP 页面符合 JSP 语法规则，即能够成为一个 JSP 页面文件。比如，当前 JSP 页面已经使用 page 指令设置了 contentType 属性的值：

   `<%@ page contentType="text/html;Charset=GB2312" %>`

若被嵌入的文件也是一个 JSP 页面，其中也使用 page 指令为 contentType 属性设置了值：

   `<%@ page contentType="application/msword" %>`

那么，合并后的 JSP 页面就两次使用 page 指令为 contentType 属性设置了不同的属性值，导致语法错误，因为 JSP 页面中的 page 指令只能为 contentType 指定一个值。

Tomcat 5.0 版本以后的服务器每次都要检查被嵌入的文件是否被修改过。因此，JSP 页面成功静态嵌入一个文件后，如果对嵌入的文件进行了修改，那么 Tomcat 服务器会重新编译 JSP 页面，即将当前的 JSP 页面和修改后的文件合并成一个 JSP 页面，然后 Tomcat 服务器将这个新的 JSP 页面转译成 Java 类文件。

【例 3-2】 在 JSP 页面静态嵌入一个文本文件 Hello.txt，该文本文件的内容是："祝贺深圳成功举办大运会"。文本文件和当前 JSP 页面在同一 Web 服务目录中。

example3_2.jsp
```
<%@ page contentType="text/html;Charset=GB2312" %>
<HTML><BODY bgcolor=yellow>
<H1>
   <%@ include file="Hello.txt" %>
</H1>
</BODY></HTML>
```

上述 example3_2.jsp 等价于下面的 JSP 文件 exm.jsp。

exm.jsp
```
<%@ page contentType="text/html;Charset=GB2312" %>
<HTML><BODY bgcolor=yellow>
<H1>祝贺深圳成功举办大运会</H1>
</BODY></HTML>
```

注： 在 Tomcat 4 版本中，不允许在 include 指令标记嵌入的文件中使用 page 指令指定 contentType 属性的值。但是在 Tomcat 5 后的版本中，在 include 指令标记嵌入的文件中必须使用 page 指令指定 contentType 属性的值，而且指定的值必须与嵌入该文件的 JSP 页面中的 page 指令指定的 contentType 属性的值相同。

3.3 动作标记 include

动作标记是一种特殊的标记，影响 JSP 运行时的行为。include 动作标记的语法有以下两种格式：

  `<jsp:include  page="文件的 URL" />`

或

  `<jsp:include  page="文件的 URL" >`
   子标记
  `</jsp:include>`

include 动作标记在 JSP 页面执行阶段处理所需要的文件，因此 JSP 页面和它所需要的文件在逻辑和语法上是独立的。注意：当 include 动作标记不需要子标记时，必须使用上述第一种形式。

尽管 include 动作标记和 include 指令标记的作用都是处理所需文件，但是在处理方式和处理时间上是不同的。include 指令标记用嵌入方式处理文件，在编译阶段就处理所需文件，优点是页面的执行速度快；而 include 动作标记不使用嵌入方式来处理所需文件，在 JSP 页面运行时才处理文件，也就是说，当 Tomcat 服务器把 JSP 页面转译成 Java 文件时，不把 JSP 页面中 include 动作标记所包含的文件与原 JSP 页面合并一个新的 JSP 页面，而是告诉 Java 解释器，这个文件在 JSP 运行时（Java 文件的字节码文件被加载执行）才被处理，其优点是能更灵活地处理所需文件（见后面的 param 标记），缺点是执行速度要慢一些。include 动作标记要处理的文件如果不是 JSP 文件，就将文件的内容发送到客户机，由客户机负责执行并显示；如果包含的文件是 JSP 文件，那么 Tomcat 服务器就执行这个 JSP 文件，然后将执行的结果发送到客户机，并由客户机负责显示这些结果。

图 3.4 使用 include 动作标记

注：书写 include 动作标记时要注意"jsp"、"："、"include"三者之间不要有空格。

【例 3-3】 example3_3.jsp 页面包含两个文件：time.jsp 和 sun.jsp。time.jsp 和 example3_3.jsp 页面在同一 Web 服务目录中，sun.jsp 在当前 Web 服务目录的子目录 image 中。（效果如图 3.4 所示）

example3_3.jsp
```
<%@ page contentType="text/html;Charset=GB2312" %>
<HTML><BODY bgcolor=red>
  <jsp:include page="time.jsp" />
  <P>请看一幅图像：<BR/>
  <jsp:include page="image/sun.jsp" />
</BODY></HTML>
```

time.jsp
```
<%@ page contentType="text/html;Charset=GB2312" %>
<%@ page import="java.util.Date" %>
<HTML><BODY><FONT size=3>
<P>现在的时间：
  <%  Date date=new Date();
      out.println(date.toString());
  %>
</FONT></BODY></HTML>
```

sun.jsp
```
<%@ page contentType="text/html;Charset=GB2312" %>
<HTML><BODY>
<IMAGE src="image/ok.jpg" width=120 height=120 >太阳图片</IMAGE>
</BODY></HTML>
```

注：sun.jsp 页面使用的图片文件 ok.jpg 需保存在当前 Web 服务目录（chaper3）的 image 子目录中。

3.4 动作标记 param

param 动作标记可以作为 include、forward 动作标记的子标记来使用，该标记以"名字-值"对的形式为这些动作标记提供附加信息。

param 动作标记的格式如下：

   <jsp:param name="名字" value="指定给 name 属性的值" />

当使用 include 动作标记动态处理所需要的 JSP 文件时，经常会使用 param 子标记，以便向动态加载的 JSP 页面传递必要的值，这也体现出 include 动作标记比 include 指令标记更灵活的特点。include 动作标记所要加载的 JSP 文件可以使用 Tomcat 服务器提供的 request 内置对象获取 param 标记中 name 属性所提供的值。

【例 3-4】 制作页面 example3_4.jsp，使用 include 动作标记动态加载 computer.jsp 文件。当 computer.jsp 文件被加载时，该文件获取 include 动作标记的 param 子标记中属性 name 的值（效果如图 3.5 所示）。

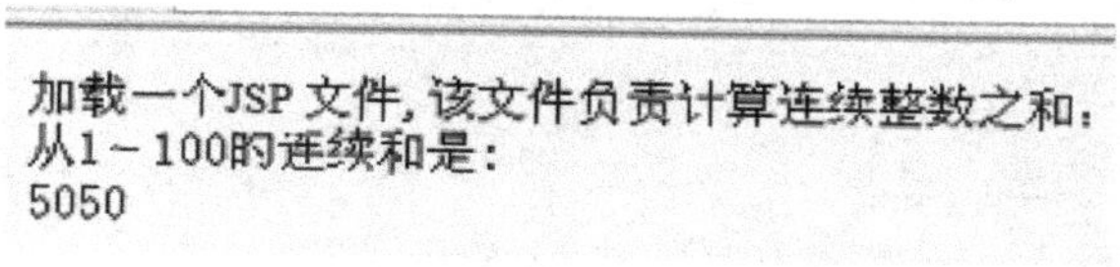

```
加载一个JSP文件,该文件负责计算连续整数之和:
从1～100的连续和是:
5050
```

图 3.5 使用 param 动作标记

example3_4.jsp
```jsp
<%@ page contentType="text/html;Charset=GB2312" %>
<HTML><BODY bgcolor=yellow>
<P>加载一个 JSP 文件，该文件负责计算连续整数之和：
    <jsp:include page="computer.jsp">
        <jsp:param name="item" value="100" />
    </jsp:include>
</BODY></HTML>
```

computer.jsp
```jsp
<%@ page contentType="text/html;Charset=GB2312" %>
<HTML><BODY>
<%  String str=request.getParameter("item");     //获取 param 标记中 name 属性的值
    int n=Integer.parseInt(str);
    int sum=0;
    for(int i=1;i<=n;i++){
        sum=sum+i;
    }
    out.println("<BR>从 1～"+n+"的连续和是： </BR>"+sum);
%>
</BODY></HTML>
```

3.5 动作标记 forward

forward 动作标记的作用是，从该标记出现处停止当前 JSP 页面的继续执行，而转向执行 forward 动作标记中 page 属性所指定的 JSP 页面。

forward 动作标记有两种格式：
    <jsp:forward page="要转向的页面" />

或

    <jsp:forward page="要转向的页面" >
        param 子标记
    </jsp:forward>

需要注意的是，当 forward 动作标记不需要 param 子标记时，必须使用上述第一种形式。forward 标记可以使用 param 动作标记作为子标记，以便向要转向的页面传送信息。forward 动作标记指定的要转向的 JSP 文件，可以使用 Tomcat 服务器提供的 request 内置对象获取 param 子标记中 name 属性所提供的值。

【例 3-5】 制作 example3_5.jsp 页面，使用 forward 标记转向 num1.jsp 或 num2.jsp 页面。在 example3_5.jsp 页面中随机获取一个 1~100 之间的整数，该数大于 50 就转向页面 num1.jsp，否则转向页面 num2.jsp。forward 动作标记使用 param 子标记，将 example3-5.jsp 页面获取的随机数传递给要转向的页面（效果如图 3.6 所示）。

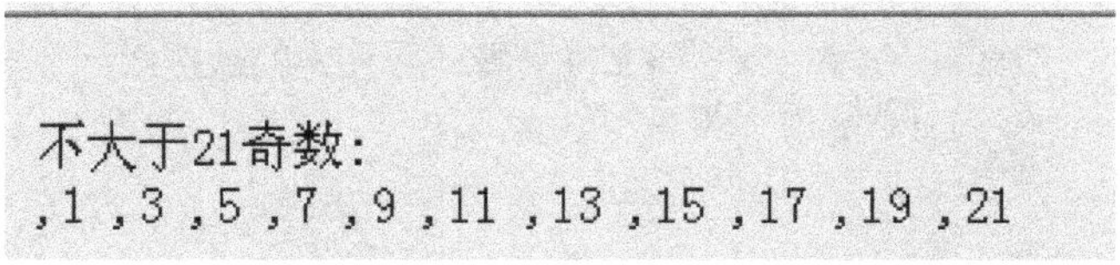

> 不大于21奇数：
> ,1 ,3 ,5 ,7 ,9 ,11 ,13 ,15 ,17 ,19 ,21

图 3.6 使用 forward 动作标记

example3_5.jsp
```jsp
<%@ page contentType="text/html;Charset=GB2312" %>
<HTML><BODY >
   <% out.println("根据不同的值转向不同的页面:<BR>");
      int n=(int)(Math.random()*100)+1;
      if(n>50) {
   %>   <jsp:forward page="num1.jsp" >
           <jsp:param name="item" value="<%= n %>" />
        </jsp:forward>
   <%
      }
      else{
   %>   <jsp:forward page="num2.jsp" >
           <jsp:param name="item" value="<%= n %>" />
        </jsp:forward>
   <%
      }
      out.println("看不见这句话");
   %>
</BODY></HTML>
```

num1.jsp
```jsp
<%@ page contentType="text/html;Charset=GB2312" %>
<HTML><BODY bgcolor=cyan >
```

```
            <%  String str=request.getParameter("item");      //获取值 param 标记中 name 属性的值
                int n=Integer.parseInt(str);
                out.println("<BR>不大于"+n+"的素数:</BR>");
                int i=0,j=0;
                for(i=1;i<=n;i++){
                    for(j=2;j<i;j++){
                        if(i%j==0)
                            break;
                    }
                    if(j==i)
                        out.println(","+i);
                }
            %>
        </BODY></HTML>
num2.jsp
        <%@ page contentType="text/html;charset=GB2312" %>
        <HTML><BODY bgcolor=cyan >
            <%  String str=request.getParameter("item");      //获取值 param 标记中 name 属性的值
                int n=Integer.parseInt(str);
                out.println("<BR>不大于"+n+"的奇数:</BR>");
                for(int i=1;i<=n;i++){
                    if(i%2!=0)
                        out.println(","+i);
                }
            %>
        </BODY></HTML>
```

3.6 动作标记 useBean

该标记用来创建并使用一个 JavaBean，它是非常重要的一个动作标记，将在第 5 章详细讨论。Sun 公司倡导的是：用 HTML 完成 JSP 页面的静态部分，用 JavaBean 完成动态部分，实现真正意义上的静态与动态的分离。

3.7 Tag 文件与 Tag 标记

一个 Web 应用中的许多 JSP 页面可能需要使用某些相同的信息，如都需要使用相同的导航栏、标题等。如果能将许多页面都需要的共同的信息形成一种特殊文件，而且各 JSP 页面都可以使用这种特殊的文件，那么这样的特殊文件就是可复用的代码。代码复用是软件设计的一个重要方面，代码复用是衡量软件可维护性的重要指标之一。为了更好地维护一个 Web 应用，JSP 页面可以通过自定义标记使用一种特殊的文件：Tag 文件（标记文件），在设计 Web 应用时，可以通过编写 Tab 文件来实现代码复用。

1. Tag 文件的结构

Tag 文件是扩展名为 .tag 的文本文件，其结构几乎与 JSP 文件相同，只是扩展名不同。

也就是说，一个 Tag 文件中可以有普通的 HTML 标记符、某些特殊的指令标记、成员变量和方法、Java 程序片和 Java 表达式等。以下是一个简单的 Tab 文件：

FirstTag.tag
```
<HTML><BODY>
  <p>这是一个 Tab 文件，负责计算 1～100 的连续和
  <%   int sum=0,i=1;
        for(i=1;i<=100;i++){
           sum=sum+i;
        }
       out.println(sum);
  %>
</BODY></HTML>
```

2. Tag 标记

当我们编写了一个 Tag 文件后，也就自定义出了一个标记，该标记的格式如下：

    `<Tag 文件名字 />`

或

    `<Tag 文件名字 >`
        标记体
    `</ Tag 文件名字>`

也就是说，一个 Tag 文件对应一个标记，习惯上称为 Tag 标记。若干 Tag 标记组成一个标记库，习惯上称为自定义标记库。

3. Tag 标记的使用

Tag 文件只供 JSP 页面使用，用户不能通过浏览器直接请求一个 Tag 文件。一个 JSP 页面使用 Tag 标记来调用相应的 Tag 文件。

为了能让一个 Web 应用中的 JSP 页面使用某些 Tag 文件，必须把这些 Tag 文件存放到 Tomcat 服务器指定的目录中，即存放到"Web 服务目录\WEB-INF\tags"中。其中的 WEB-INF 和 tags 都是固定的子目录名称，而 tags 下的子目录名字可由用户给定。也就是说，tags 或其下的子目录是专门存放 Tag 文件的，而一个 Tag 文件对应一个 Tag 标记，因此习惯上称 tags 或其下的子目录为一个标记库。

一个 JSP 页面必须使用<taglib>指令标记引入标记库，只有这样，JSP 页面才可以使用 Tag 标记调用相应的 Tag 文件。<taglib>指令的格式如下：

    `<%@ taglib tagdir="自定义标记库的位置"  prefix="前缀">`

一个 JSP 页面可以使用几个<taglib>指令标记引入若干个标记库。引入标记库后，JSP 页面就可以使用带前缀的 Tag 标记调用相应的 Tag 文件，其中的前缀由<taglib>指令中的 prefix 属性指定。例如：

    `<前缀：Tag 文件名字/>`

前缀可以有效地区分不同标记库中具有相同名字的标记文件。

当使用 JSP 页面调用一个 Tag 文件时，Tomcat 服务器负责处理相应的 Tag 文件，也就是说，Tamcat 服务器将 Tag 文件转译成一个 Java 文件，再将这个 Java 文件编译生成字节码文件，并将这个字节码文件加载到内存中，然后执行这个字节码文件，就好像它是一个常规的

JSP 文件一样。

现在用一个例子来说明怎样在 JSP 页面中调用 Tag 文件。假设使用的 Web 服务目录是 Tomcat 安装目录 webapps 下的一个子目录 chapter3。我们编写 FirstTag.tag 和 SecondTag.tag 两个 Tag 文件，其中 FirstTag.tag 文件的内容参见 3.7.1 节，SecondTag.tag 文件的内容如下：

SecondTag.tag
```
<p>这是一个 Tab 文件负责计算 20 内的素数：
    <%  int i,j;
        for(i=1;i<=20;i++){
            for(j=2;j<i;j++){
                if(i%j==0)
                    break;
            }
            if(j==i)
                out.println("<BR>素数:"+i);
        }
    %>
```

在 chapter3 下建立目录结构：chapter3\WEB-INF\tags\tagsTwo，将 FirstTag.tag 存放在 tags 目录中，而 SecondTag.tag 存放在 tags 的子目录 tagsTwo 中。

【例 3-6】 页面文件 example3_6.jsp 使用了 FirstTag.tag 和 SecondTag.tag 文件（效果如图 3.7 所示）。

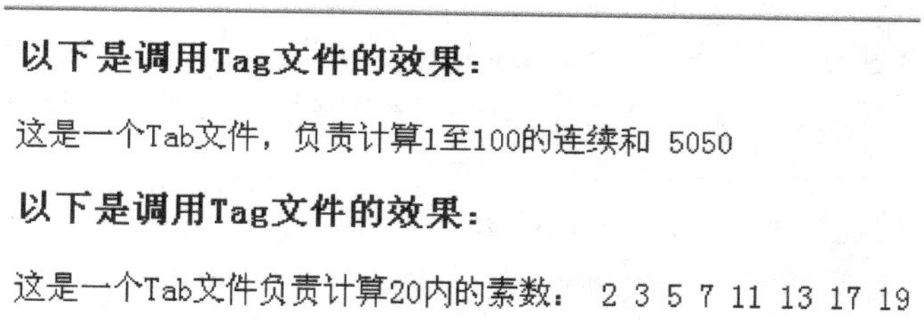

以下是调用Tag文件的效果：

这是一个Tab文件，负责计算1至100的连续和 5050

以下是调用Tag文件的效果：

这是一个Tab文件负责计算20内的素数： 2 3 5 7 11 13 17 19

图 3.7　使用 Tag 文件

example3_6.jsp
```
<%@ page contentType="text/html;Charset=GB2312" %>
<%@ taglib tagdir="/WEB-INF/tags" prefix="com"%>
<%@ taglib tagdir="/WEB-INF/tags/tagsTwo" prefix="game"%>
<HTML>   <BODY>
    <H3>以下是调用 Tag 文件的效果：</H3>
        <com:FirstTag />
    <H3>以下是调用 Tag 文件的效果：</H3>
        <game:SecondTag />
</BODY> </HTML>
```

4．Tag 文件中的常用指令

与 JSP 文件类似，Tag 文件中也有一些常用指令，这些指令将影响 Tag 文件的行为。Tag 文件中经常使用的指令有 tag、taglib、include、attribute、variable。以下将分别讲述上述指令在 Tag 文件中的作用和用法。

（1）tag 指令

Tag 文件中的 tag 指令类似于 JSP 文件中的 page 指令。Tag 文件通过使用 tag 指令可以指定某些属性的值，以便从总体上影响标记文件的处理和表示。tag 指令的语法如下：

    <%@ tag 属性 1="属性值" 属性 2="属性值" …属性 n="属性值" >

一个 Tag 文件中可以使用多个 tag 指令，因此经常使用多个 tag 指令为属性指定需要的值：

    <%@ tag 属性 1="属性值" >
    <%@ tag 属性 2="属性值" >
    ……
    <%@ tag 属性 n="属性值" >

tag 指令可以操作的属性有 body-content、language、import、pageEncoding。以下分别讲述怎样设置这些属性的值。

① body-content 属性

我们已经知道，一个 Tag 文件会对应一个自定义标记，该标记的格式如下：

    <Tag 文件名字 />

或

    <Tag 文件名字 >
        标记体
    </ Tag 文件名字>

JSP 文件通过使用该自定义标记调用相应的 Tag 文件，那么 JSP 文件到底应该使用自定义标记的哪种格式来调用 Tag 文件呢？

一个 Tag 文件通过 tag 指令指定 body-content 属性的值，以决定 Tag 标记的使用格式，也就是说，该属性的值可以确定 JSP 页面使用 Tag 标记时是否可以有标记体，如果允许有标记体，该属性会给出标记体内容的类型。

body-content 属性值有 empty、tagdependent、scriptless，默认值是 scriptless。

如果 body-content 属性的值是 empty，那么 JSP 页面必须使用没有标记体的 Tag 标记

    <Tag 文件名字 />

来调用相应的 Tag 文件。

如果 body-content 属性的值是 tagdependent 或 scriptless，那么 JSP 页面可以使用无标记体或有标记体的 Tag 标记

    <Tag 文件名字 >
        标记体
    </ Tag 文件名字>

来调用相应的 Tag 文件。如果属性值是 scriptless，那么标记体中不能有 Java 程序片；如果属性值是 tagdependent，那么标记体的内容将按纯文本处理。

Tag 标记中的标记体由相应的 Tag 文件负责处理，因此，当 JSP 页面使用有标记体的 Tag 标记调用一个 Tag 文件时，可以通过"标记体"向该 Tag 文件动态地传递文本数据或必要的 JSP 指令。

Tag 文件通过使用指令

    <jsp:doBody />

来获得 JSP 页面传递过来的"标记体"，也就是说，在一个 Tag 文件中，<jsp:doBody/>指令被替换成"标记体"的内容。

【例 3-7】 Show.tag 文件负责用某种颜色显示文本，example3_7.jsp 页面负责调用 Show.tag 文件，而且通过标记体向 Show.tag 文件传递文本内容（效果如图 3.8 所示）。

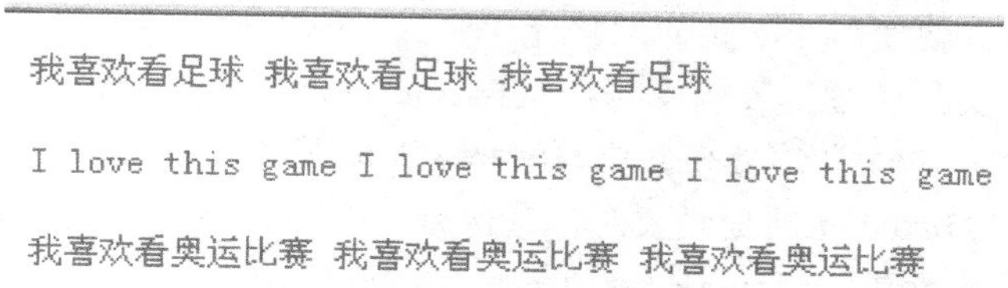

我喜欢看足球　我喜欢看足球　我喜欢看足球

I love this game I love this game I love this game

我喜欢看奥运比赛　我喜欢看奥运比赛　我喜欢看奥运比赛

图 3.8　使用带标记体的 Tag 标记

example3_7.jsp

```
<%@ page contentType="text/html;Charset=GB2312" %>
<%@ taglib prefix="tags" tagdir="/WEB-INF/tags" %>
<HTML>
    <tags:Show>
      我喜欢看足球
    </tags:Show>
     <tags:Show>
       I love this game
    </tags:Show>
    <tags:Show>
      我喜欢看奥运比赛
    </tags:Show>
</HTML>
```

Show.tag

```
<BODY> <P>
<% for(int i=1;i<=3;i++){    %>
    <FONT color="red" size="3">
       <jsp:doBody />
    </FONT>
<% }
%>
</P></BODY>
```

② language 属性

language 属性的值指定 Tag 文件使用的脚本语言，目前只能取值 Java，其默认值就是 Java，因此在编写 Tag 文件时，没有必要使用 tag 指令指定 language 属性的值。

③ import 属性

import 属性的作用是为 Tag 文件引入 Java 核心包中的类，这样就可以在 Tag 文件的程序片部分、变量及函数声明部分、表达式部分使用包中的类。import 属性可以取多个值，import 属性已经有如下值："java.lang.*"、"javax.servlet.*"、"javax.servlet.jsp.*"、"javax.servlet.http.*"。

④ pageEncoding 属性

该属性指定 Tag 文件的字符编码，其默认值是 ISO-8859-1。Tag 文件必须使用 ANSI 编码保存。

（2）include 指令

在 Tag 文件中也可以使用 include 指令标记，其使用方法和作用与 JSP 文件中类似。

（3）attribute 指令

Tag 文件充当着可复用代码的角色，如果一个 Tag 文件能允许使用它的 JSP 页面向该 Tag 文件传递字符串数据，就使得 Tag 文件的功能更强大。在 Tag 文件中通过使用 attribute 指令，可以动态地向该 Tag 文件传递需要的字符串数据。attribute 指令的格式如下：

<%@ attribute name="字符串变量名" required="true"|"false"%>

attribute 指令中的 name 属性是必要的，其值是一个字符串变量的名字。JSP 页面在使用 Tag 文件时，可向 Tag 文件中 attribute 指令中的 name 属性指定的字符串变量传递一个字符串。

JSP 页面使用 Tag 标记向调用的 Tag 文件中 name 属性指定的字符串变量传递一个字符串，方式如下：

<前缀：Tag 文件名字 字符串变量名= "字符串" />

或

<前缀：Tag 文件名字 字符串变量名= "字符串" >
　　标记体
</前缀：Tag 文件名字>

attribute 指令中的 required 属性是可选的，如果省略 required 属性，那么 required 的默认值是 false。当指定 required 的值是 true 时，使用该 Tag 文件的 JSP 页面必须向该 Tag 文件中 attribute 指令中的 name 属性指定的字符串变量传递值，即当 required 的值是 true 时，如果使用"<前缀:Tag 文件名字 />"调用 Tag 文件就会出现错误。当指定 required 的值是 false 时，使用该 Tag 文件的 JSP 可以向该 Tag 文件中 attribute 指令中的 name 属性指定的字符串变量传递或不传递值。

【例 3-8】 Tag 文件 Trangle.tag 负责计算显示三角形的面积。example3_8.jsp 页面在调用 Trangle.tag 文件时向 Trangle.tag 文件传递三角形三边的长度。example3_8.jsp 页面存放在 Web 服务目录 chapter3 中，Trangle.tag 存放在 chapter3\WEB-INF\tags 中（效果如图 3.9 所示）。

```
以下是调用 Tag 文件的效果：
这是一个 Tag 文件，负责计算三角形的面积。
JSP 页面传递过来的三条边：3，4，5
三角形的面积为：
```

图 3.9　向 Tag 文件传递数据

example3_8.jsp
```
<%@ page contentType="text/html;Charset=GB2312" %>
<%@ taglib tagdir="/WEB-INF/tags" prefix="computer"%>
<HTML><BODY>
    <H3>以下是调用 Tag 文件的效果：</H3>
    <computer:Trangle sideA="3" sideB="4" sideC="5"/>
</BODY></HTML>
```

Trangle.tag
```
<p>这是一个 Tag 文件，负责计算三角形的面积。
<%@ attribute name="sideA" required="true" %>
<%@ attribute name="sideB" required="true" %>
<%@ attribute name="sideC" required="true" %>
```

```
<%!   public String getArea(double a, double b, double c) {
          if(a+b>c&&a+c>b&&c+b>a) {
              double p=(a+b+c)/2.0;
              double area=Math.sqrt(p*(p-a)*(p-b)*(p-c)) ;
              return "<BR>三角形的面积为："+area;
          }
          else{
              return("<BR>"+a+","+b+","+c+"不能构成一个三角形，无法计算面积");
          }
      }
%>
<%    out.println("<BR>JSP 页面传递过来的三条边："+sideA+","+sideB+","+sideC);
      double a=Double.parseDouble(sideA);
      double b=Double.parseDouble(sideB);
      double c=Double.parseDouble(sideC);
      out.println(getArea(a,b,c));
%>
```

习题 3

1. JSP 页面使用 page 指令可以为属性 contentType 指定几个值？
2. JSP 页面使用 page 指令可以为 import 属性指定多个值吗？
3. include 指令标记和 include 动作标记有何不同？
4. param 动作标记经常作为哪些标记的子标记，有何作用？
5. Tag 文件应当存放在怎样的目录中？
6. 用户可以使用浏览器直接访问一个 Tag 文件吗？
7. JSP 页面怎样调用一个 Tag 文件？
8. 编写两个 Tag 文件 Rect.tag 和 Circle.tag。Rect.tag 负责计算矩形的面积，Circle.tag 负责计算圆的面积。编写一个 JSP 页面，该 JSP 页面使用 Tag 标记调用 Rect.tag 和 Circle.tag。调用 Rect.tag 时，向其传递矩形的两个边的长度；调用 Circle.tag 时，向其传递圆的半径。

第 4 章

内 置 对 象

> **本章导读**
> ✿ 知识点：掌握 JSP 内置对象：resquest、response、session、application、out 的作用以及使用方法。
> ✿ 重点：理解 request、response 和 session 对象在 Web 设计中的重要性。
> ✿ 难点：学习使用 session 对象，理解 session 对象的生命周期。
> ✿ 关键实践：编写 JSP 页面，使用 session 对象存储有关数据。

一个 JSP 页面不仅可以有 HTML 标记和 JSP 标记，而且可以有成员变量声明，JSP 页面可以在 Java 程序片和 Java 表达式中使用 JSP 页面声明的成员变量。有些成员变量不用声明就可以在 JSP 页面的脚本（Java 程序片和 Java 表达式）中使用，这就是所谓的内置对象。

内置对象有 resquest、response、session、application、out，以下将一一介绍。

本章将在目录 webapps 下新建一个 Web 服务目录 chapter4。除非特别约定，本章例子中涉及的 JSP 页面均保存在 chapter4 中，Tag 文件均保存在 chapter4\WEB-INF\tags 中。

4.1 request 对象

为了更好地理解 request 对象，需要了解 HTTP（HyperText Transfer Protocol，超文本传输协议）。HTTP 是客户与服务器之间一种请求（request）信息与响应信息（response）的通信协议。我们使用浏览器从网站获取 HTML 页面或 JSP 页面时，遵守的就是 HTTP。HTTP 规定了信息在 Internet 上的传输方法，特别规定了浏览器与服务器的交互方法。

从网站获取页面时，浏览器在网站上打开了一个对网络服务器的连接，并发出请求。服务器收到请求并做出响应，响应结束后（用户浏览器显示了页面的效果），客户浏览器和服务器之间的连接关闭。所以，HTTP 被称为"请求和响应"协议。

用户通过在浏览器的地址栏中输入一个 HTML 或 JSP 页面所在的服务器的地址和页面的

名字来请求该页面。当用户通过浏览器请求一个页面时，浏览器将请求发送给服务器。按照HTTP，浏览器发送的请求具有一定的结构，请求包括一个请求行、头域和表单提交的信息体等。服务器会根据浏览器发送来的请求做出响应，并将相应结果返回给用户的浏览器。最简单的请求是对页面的一个简单请求，例如：

GET/hello.html HTTP/1.1
host:www.sina.com.cn

上面的第一行称为请求行，含有请求使用的方法、请求的资源和请求使用的协议；第二行称为头（header），给出请求的主机地址，其中请求的方法是 get 方法（其他方法还包括 post、head、delete、trace 及 put 方法等），请求的资源是 hello.html，使用 HTTP 1.1 版本，请求的主机是 www.sina.com.cn。

一个典型请求通常包含许多头，称为请求的 HTTP 头，提供了关于信息体的附加信息及请求的来源。其中有些头是标准的，有些头与特定的浏览器有关。

一个请求还可能包含信息体，如信息体可包含 HTML 表单的内容。在 HTML 表单上单击【submit】按钮时，如果表单使用 ACTION="POST"或 ACTION="GET"特征，请求将使用 post 方法或 get 方法将表单中输入的信息发送给服务器。

1．获取客户提交的信息

当用户请求一个 JSP 页面时，JSP 页面所在的 Tomcat 服务器将用户的请求封装在内置对象 request 中。那么，该对象调用相应的方法可以获取封装的信息，也就是说，使用该对象可以获取用户浏览器提交的请求信息，以便做出相应的响应。尽管 request 对象有许多可调用的方法，但是 request 对象常用的方法是 getParameter(String s)，该方法获取表单提交的信息。

内置对象 request 对象是实现了 ServletRequest 接口类的一个实例，可以在 Tomcat 服务器的 webapps\tomcat-docs\servletapi 目录中查找 ServletRequest 接口的方法。

【例 4-1】用户可以使用 example4_1.jsp 提供的表单再次请求 example4_1.jsp 页面，可以在表单提供的文本框中输入一个数字，并提交给 example4_1.jsp 页面，该页面通过内置对象获取用户提交的数字，然后让一个 Tag 文件负责计算该数字的平方，并将计算结果返回给用户（效果如图 4.1 所示）。

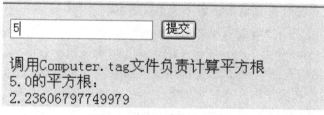

图 4.1 获取表单提交的信息

example4_1.jsp

```
<%@ page contentType="text/html;Charset=GB2312" %>
<%@ taglib tagdir="/WEB-INF/tags" prefix="com"%>
<HTML><BODY bgcolor=cyan><FONT size=3>
  <FORM   action="" method=post name=form>
    <Input type="text" name="number">
    <Input type="submit" value="提交" name="submit">
```

```
        </FORM>
        <%    String textContent=request.getParameter("number");
              if(textContent==null) {
                  out.println("请在文本框中输入数字，按提交按钮");
              }
              else{
        %>        <p>调用 Computer.tag 文件负责计算平方根
                  <com:Computer number="<%= textContent %>" />
        <%    }
        %>
        </FONT></BODY></HTML>
```

Compter.tag
```
        <%@ attribute name="number" %>
        <%    try{
                  double n=Double.parseDouble(number);
                  if(n>=0) {
                      double r=Math.sqrt(n) ;
                      out.print("<BR>"+n+"的平方根： ");
                      out.print("<BR>"+r);
                  }
                  else
                      out.print("<BR>"+"请输入一个正数");
              }
              catch(NumberFormatException e) { out.print("<BR>"+"请输入数字字符"); }
        %>
```

在运行例 4-1 时，用户先在浏览器的地址栏中输入 example4_1.jsp 页面地址"http://127.0.0.1:8080/chaper4/example4_1.jsp"，服务器做出响应，将表单发送给用户的浏览器（用户就会看到 example4_1.jsp 页面的表单），服务器的内置对象 request 调用 getPeremeter()方法获取用户通过表单的文本框提交的信息。但是由于用户首次请求 example4_1.jsp 页面，还没有在表单中输入信息，没有单击表单中的【提交】按钮，所以 request 调用 getPeremeter()方法获取信息为 null。当用户看到表单之后，就可以在表单的文本框中输入信息，单击【提交】按钮，再次请求 example4_1.jsp（表单中的 action 指定该表单要请求的页面，如果是当前页面，可以用""代替当前页面），那么服务器再次做出响应。这时，request 调用 getPeremeter()方法，就可以获取用户在表单的文本框中输入的信息（如果用户没有在文本框输入任何信息就单击表单中的【提交】按钮，getPeremeter()方法获取的字符串长度为 0，即该字符串为""）。

2．处理汉字信息

当用 request 对象获取请求中的汉字字符信息时，可能出现乱码问题，所以对含有汉字字符的信息必须进行特殊的处理方式。避免出现汉字乱码问题有两种方式，一种方式在第 3 章中讲述 page 指令提到过，即在使用 page 指定 contentType 属性的值时进行如下指定：

```
<%@ page contentType="text/html;Charset=GB2312" %>
```

将其中出现的"Charset"中的首写字母大写（C 为大写字母）。另一种方式是将

```
<%@ page contentType="text/html;charset=GB2312" %>
```
中出现的"charset"中的首写字母小写（c 为小写字母），内置对象将获取信息重新编码，即用 ISO-8859-1 进行编码，并把编码存放到一个字节数组中，再把这个数组转化为字符串，如下所示：

```
String str=request.getParameter("girl");
byte   b[]=str.getBytes("ISO-8859-1");
str=new String(b);
```

【例 4-2】 页面文件 example4_2.jsp 通过两个表单分别向页面 showMessage1.jsp 和 showMessage2.jsp 提交信息，showMessage1.jsp 和 showMessage2.jsp 负责显示用户提交的信息，并分别使用上述两种不同的方式来避免汉字出现乱码。

example4_2.jsp
```
<%@ page contentType="text/html;Charset=GB2312" %>
<HTML><BODY bgcolor=cyan>
   <FORM action="showMessage1.jsp" method=post name=form>
      <Input type="text" name="boy">
      <Input TYPE="submit" value="提交给 showMessage1.jsp" name="submit">
   </FORM>
   <FORM action="showMessage2.jsp" method=post name=form>
      <Input type="text" name="boy">
      <Input TYPE="submit" value="提交给 showMessage2.jsp" name="submit">
   </FORM>
</BODY></HTML>
```

showMessage1.jsp
```
<%@ page contentType="text/html;Charset=GB2312" %>
<HTML><BODY>
<P>获取文本框提交的信息：
   <%   String textContent=request.getParameter("boy");
   %>
<BR>
   <%=textContent%>
<P> 获取按钮的名字：
   <%   String buttonName=request.getParameter("submit");
   %>
<BR>   <%=buttonName%>
</BODY></HTML>
```

showMessage2.jsp
```
<%@ page contentType="text/html;charset=GB2312" %>
<HTML ><BODY>
<P>获取文本框提交的信息：
   <%   String textContent=request.getParameter("boy");
        byte   b[]=textContent.getBytes("ISO-8859-1");
        textContent=new String(b);
   %>
<BR>   <%=textContent%>
<P> 获取按钮的名字：
```

```
<%  String buttonName=request.getParameter("submit");
    byte   c[]=buttonName.getBytes("ISO-8859-1");
    buttonName=new String(c);
%>
<BR> <%=buttonName%>
</BODY></HTML>
```

3. request 对象的常用方法

我们已经知道，当用户访问一个 JSP 页面时，会提交一个 HTTP 请求给 Tomcat 服务器，这个请求包括一个请求行、HTTP 头和信息体。例如：

```
POST/ E.jsp/HTTP.1.1
host: localhost: 8080
accept-encoding：gzip, deflate
```

其中，首行叫做请求行，规定了向访问的页面请求提交信息的方式，如 post、get 等方法，以及请求的页面的文件名字和使用的通信协议。

第 2、3 行分别是两个头（header），host、accept-encoding 称为头名字，localhost:8080 和 gzip、deflate 分别是它们的值。这里，host 的值是 E.jsp 的地址。上面的请求有两个头：host 和 accept-encoding。一个典型的请求通常包含很多头，有些头是标准的，有些头与特定的浏览器有关。

尽管服务器非常关心用户提交的 HTTP 请求中表单的信息，如例 4-1 和例 4-2 中使用 request 的 getParameter()方法获取表单提交的有关信息，但实际上，使用 request 对象调用相关方法可以获取请求的许多细节信息。内置对象 request 常用方法如下。

- getProtocol()：获取请求使用的通信协议，如 http/1.1 等。
- getServletPath()：获取请求的 JSP 页面所在的目录。
- getContentLength()：获取 HTTP 请求的长度。
- getMethod()：获取表单提交信息的方式，如 post 或 get。
- getHeader(String s)：获取请求中头的值。一般，s 参数可取的头名有 accept、referrer、accept-language、content-type、accept-encoding、user-agent、host、content-length、connection、cookie 等。比如，s 取值 user-agent 将获取客户的浏览器的版本号等信息。
- getHeaderNames()：获取头名字的一个枚举。
- getHeaders(String s)：获取头的全部值的一个枚举。
- getRemoteAddr()：获取客户的 IP 地址。
- getRemoteHost()：获取客户机的名称（如果获取不到，就获取 IP 地址）。
- getServerName()：获取服务器的名称。
- getServerPort()：获取服务器的端口号。
- getParameterNames()：获取表单提交的信息体部分中 name 参数值的一个枚举。

【例 4-3】 使用 request 的一些常用方法。

example4_3.jsp

```
<%@ page contentType="text/html;Charset=GB2312" %>
<%@ page import="java.util.*" %>
<HTML><BODY bgcolor=cyan><Font size=3>
```

```
<FORM action="" method=post name=form>
    <Input type="text" name="boy">
    <Input TYPE="submit" value="enter" name="submit">
</FORM>
<TABLE border=1>
    <% String protocol=request.getProtocol();
       String path=request.getServletPath();
       String method=request.getMethod();
       String header=request.getHeader("accept");
    %>
    <TR>
        <TD>客户使用的协议是：</TD>
        <TD>"<%= protocol %>"</TD>
    </TR>
    <TR>
        <TD>用户请求的页面所在位置：</TD>
        <TD>"<%= path %>"</TD>
    </TR>
    <TR>
        <TD>客户提交信息的方式：</TD>
        <TD>"<%= method %>"</TD>
    </TR>
    <TR>
        <TD>>获取 HTTP 头文件中 accept 的值（浏览器支持的 MIME 类型）：</TD>
        <TD>"<%= header %>"</TD>
    </TR>
</TABLE>
<BR>获取客户端提交的所有参数的名字：
    <%  Enumeration enumName=request.getParameterNames();
        while(enumName.hasMoreElements()) {
            String s=(String)enumName.nextElement();
            out.println(s);
        }
    %>
<BR>获取头名字的一个枚举：
    <%  Enumeration enumHeaded=request.getHeaderNames();
        while(enumHeaded.hasMoreElements()) {
            String s=(String)enumHeaded.nextElement();
            out.println(s);
        }
    %>
<BR>获取头文件中指定头名字的全部值的一个枚举：
    <%  Enumeration enumHeadedValues=request.getHeaders("cookie");
        while(enumHeadedValues.hasMoreElements()) {
            String s=(String)enumHeadedValues.nextElement();
            out.println(s);
        }
```

```
            %>
        <P> 文本框 text 提交的信息：
            <%  String textContent=request.getParameter("boy");
                if(textContent==null) {
                    textContent="";
                }
                int length=textContent.length();
                out.println(textContent);
                out.println("文本框中字符的个数："+length);
            %>
        </FONT></BODY></HTML>
```
使用 request 对象获取信息时要格外小心，在例 4-3 中：

  String textContent =request.getParameter("boy");

获取提交的字符串信息，并且在下面的代码中使用了这个字符串对象（对象调用方法）：

  length= textContent.length();

  那么，Tomcat 服务器在运行这个 JSP 页面生成的字节码文件时，会认为已使用了空对象，因为在这个字节码被执行时（客户请求页面时），客户机可能还没有提交数据（单击提交键），所以 textContent 对象还没有被创建。如果使用了空对象，即还没有创建对象就使用了该对象，Java 解释器就会提示 NullPointerException 异常。当然，如果不使用空对象就不会出现异常。

  因此可以像例 4-3 那样，为了避免在运行时 Tomcat 服务器认为已使用了空对象，使用如下代码：

```
        String textContent=request.getParameter("boy");
        if(textContent==null) {
            textContent="";
        }
```

4.2 response 对象

  当客户访问一个服务器的页面时，会提交一个 HTTP 请求，服务器收到请求后，返回 HTTP 响应。响应与请求类似，也有某种结构，每个响应都由状态行开始，可以包含几个头和可能的信息体（用户看到的页面信息）。

  request 对象获取客户请求提交的信息，与 request 对象相对应的对象是 response 内置对象。response 对象对客户的请求做出响应，向客户端发送数据。

1. 改变 contentType 属性的值

  当一个客户请求访问一个 JSP 页面时，如果该页面用 page 指令设置页面的 contentType 属性的值为 text/html，那么 response 对象按照这种属性值做出响应，将页面的静态部分返回给客户。由于 page 指令只能为 contentType 指定一个值，来决定响应的 MIME 类型，如果想动态地改变这个属性的值来响应客户，就需要让 response 对象调用 setContentType(String s) 方法来改变 contentType 的属性值：

    public void setContentType(String s)

  该方法设置响应的 MIME 类型，参数 s 可取值 text/html、text/plain、image/gif、image/jpeg、 image/x-xbitmap、image/pjpeg、application/x-shockwave-flash、application/vnd.ms-powerpoint、application/vnd.ms-excel、application/msword 等。

当服务器用 setContentType 方法动态改变了 contentType 的属性值,即响应的 MIME 类型,并将 JSP 页面的输出结果按照新的 MIME 类型返回给客户时,客户端要保证支持这种新的 MIME 类型。

【例 4-4】 客户单击不同的按钮,可以改变页面响应的 MIME 类型。当选择将当前页面用"MS-Word 显示"时,JSP 页面动态地改变 contentType 的属性值为 application/msword,客户浏览器会调用本地的 Word 软件来显示当前页面;选择将当前页面用"MS-Powerpoint 显示"时,JSP 页面动态地改变 contentType 的属性值为 application/vnd.ms-powerpoint,客户浏览器会调用本地的 PowerPoint 软件来显示当前页面。效果如图 4.2 所示。

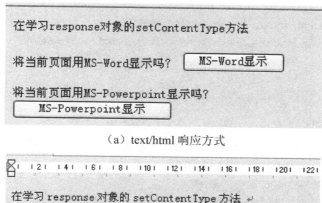

(a) text/html 响应方式

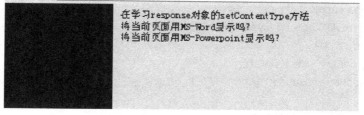

(b) application/msword 响应方式

(c) application/vnd.ms-powerpoint 响应方式

图 4.2 例 4-4 的显示效果

example4_4.jsp

```
<%@ page contentType="text/html;Charset=GB2312" %>
<HTML><BODY bgcolor=cyan><FONT size=2>
    <P>在学习 response 对象的 setContentType 方法
    <FORM action="" method="post" name=form>
```

```
            <P>将当前页面用 MS-Word 显示吗?
            <Input type="submit" value="MS-Word 显示" name="submit">
            <P>将当前页面用 MS-Powerpoint 显示吗?
            <Input type="submit" value="MS-Powerpoint 显示" name="submit">
        </FORM>
    <%  String str=request.getParameter("submit");
        if(str==null) {
            str="";
        }
        if(str.equals("MS-Word 显示")) {
            response.setContentType("application/msword");
        }
        else if(str.equals("MS-Powerpoint 显示")) {
            response.setContentType("application/vnd.ms-powerpoint");
        }
    %>
    </FONT></BODY></HTML>
```

2. 设置响应的 HTTP 头

我们已经知道,当客户访问一个页面时,会提交一个 HTTP 头给服务器。同样,响应也包括一些头。response 对象可以使用如下方法:

  addHeader(String head,String value);
  setHeader(String head,String value)

动态添加新的响应头和头的值,将这些头发送给客户机的浏览器。如果添加的头已经存在,则先前的头被覆盖。

【例 4-5】 为 response 对象添加一个响应头"refresh",其值是"5"。那么,客户机收到这个头之后,5 秒后将再次刷新页面 example4_5.jsp,导致该页面每 5 秒刷新一次。

example4_5.jsp

```
<%@ page contentType="text/html;charset=GB2312" %>
<%@ page import="java.util.*" %>
<HTML><BODY bgcolor=cyan><FONT size=4>
<P>现在的时间是: <BR>
<%  out.println(""+new Date());
    response.setHeader("Refresh","5");
%>
</FONT></BODY></HTML>
```

3. 重定向

在某些情况下,当响应客户机时,需要将客户机重新引导至另一个页面。例如,如果用户输入的表单信息不完整,就会再被引导到该表单的输入页面。

可以使用 response 的 sendRedirect(URL url)方法实现客户的重定向。

【例 4-6】 客户机在页面 example4_6.jsp 中填写表单提交给页面 form.jsp,如果填写的表单不完整,就会重新定向到页面 example4_6.jsp。

example4_6.jsp

```jsp
<%@ page contentType="text/html;charset=GB2312" %>
<HTML><BODY>
<P>填写姓名：<BR>
  <FORM action="form.jsp" method="post" name=form>
     <Input type="text" name="boy">
     <Input typE="submit" value="Enter">
  </FORM>
</BODY></HTML>
```

form.jsp

```jsp
<%@ page contentType="text/html;Charset=GB2312" %>
<HTML><BODY>
<% String str=null;
    str=request.getParameter("boy");
    if(str.length()==0) {
       response.sendRedirect("example4_6.jsp");
    }
    else{
       out.print("欢迎您来到本网页！");
       out.print(str);
    }
%>
</BODY></HTML>
```

4．状态行

当服务器对客户请求进行响应时，首先要发送的是状态行，然后发送 HTTP 头和信息体。也就是说，状态行是响应的首行。

状态行包括 3 位数字的状态码和对状态代码的描述（称为原因短语）。下面列出了 5 种状态码的大概描述。

- 1yy（1 开头的 3 位数）：主要是实验性质的。
- 2yy：表明请求成功，如状态码 200 可以表明已成功取得了请求的页面。
- 3yy：表明在请求满足之前应采取进一步的行动。
- 4yy：当浏览器无法满足请求时，返回该状态码，如 404 表示请求的页面不存在。
- 5yy：表示服务器出现问题，如 500 说明服务器内部发生错误。

一般不需要修改状态行，在出现问题时，response 对象会自动响应，发送相应的状态代码。我们也可以使用 response 对象的 setStatus(int n)方法来设置状态行，如果设置的状态行已经存在，则先前的状态行被覆盖。表 4.1 是状态码表。

【例 4-7】 使用 setStatus(int n)方法设置响应的状态行。效果如图 4.3 所示。

example4_7.jsp

```jsp
<%@ page contentType="text/html;charset=GB2312" %>
```

表 4.1 状态码表

状态码	代码说明	状态码	代码说明
101	服务器正在升级协议	404	请求的资源不可用
100	客户可以继续	405	请求所用的方法是不允许的
201	请求成功且在服务器上创建了新的资源	406	请求的资源只能用请求不能接受的内容特性来响应
202	请求已被接受但还没有处理完毕	407	客户必须得到认证
200	请求成功	408	请求超时
203	客户端给出的元信息不是发自服务器的	409	发生冲突，请求不能完成
204	请求成功，但没有新信息	410	请求的资源已经不可用
205	客户必须重置文档视图	411	请求需要一个定义的内容长度才能处理
206	服务器执行了部分 get 请求	413	请求太大，被拒绝
300	请求的资源有多种表示法	414	请求的 URL 太大
301	资源已经被永久移动到新位置	415	请求的格式被拒绝
302	资源已经被临时移动到新位置	500	服务器发生内部错误，不能服务
303	应答可以在另外一个 URL 中找到	501	不支持请求的部分功能
304	GET 方式请求不可用	502	从代理和网关接受了不合法的字符
305	请求必须通过代理来访问	503	HTTP 服务暂时不可用
400	请求有语法错误	504	服务器在等待代理服务器应答时发生超时
401	请求需要 HTTP 认证	505	不支持请求的 HTTP 版本
403	取得了请求但拒绝服务	—	—

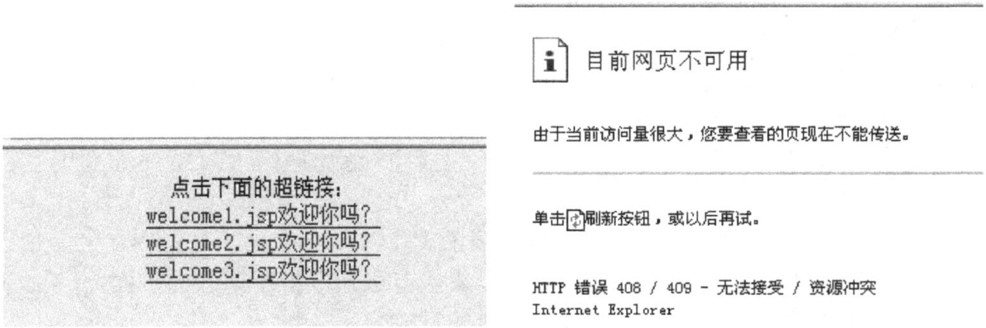

图 4.3 设置响应状态行

```
<HTML><BODY bgcolor=cyan><FONT size=2>
<P>点击下面的超链接：<BR>
    <A href="welcome1.jsp"> welcome1.jsp 欢迎你吗?
<BR> <A href=" welcome2.jsp "> welcome2.jsp 欢迎你吗?
<BR><A href=" welcome3.jsp "> welcome3.jsp 欢迎你吗?
</FONT></BODY></HTML>
```

welcome1.jsp
```
<%@ page contentType="text/html;charset=GB2312" %>
<HTML><BODY>
   <%   response.setStatus(408);
```

```
                out.print("能看到本页面吗？");
            %>
        </BODY></HTML>
welcome2.jsp
        <%@ page contentType="text/html;charset=GB2312" %>
        <HTML><BODY>
            <%  response.setStatus(200);
                out.println("这是 welcome2,能看到 welcome2.jsp 页面吗？ ");
            %>
        </BODY></HTML>
welcome3.jsp
        <%@ page contentType="text/html;charset=GB2312" %>
        <HTML><BODY>
            <%  response.setStatus(500);
            %>
        </BODY></HTML>
```

4.3 session 对象

网络通信是以协议为基础的。HTTP 是一种无状态协议，即客户机向服务器发出请求（request），然后服务器返回响应（response），连接就被关闭了，在服务器端不保留连接的有关信息。因此当下一次连接时，服务器已没有以前的连接信息了，无法判断这一次连接和以前的连接是否属于同一客户机。而且，当一个客户机访问一个服务器时，可能会在某个 Web 服务目录或其子目录中的几个页面反复连接，反复刷新一个页面，或不断地向一个页面提交信息等，服务器应当通过某种办法知道这是同一个客户机。也就是说，服务器必须使用某种手段记录有关连接的信息，这就需要 Tomcat 服务器提供的内置 session（会话）对象。session 对象在客户和服务器进行信息交互中起着非常重要的作用。

内置对象 session 由 Tomcat 服务器负责创建，session 是实现了 HttpSession 接口类的一个实例，可以在 Tomcat 服务器的 webapps\tomcat-docs\servletapi 中查找 HttpSession 接口的方法。

1. session 对象的 ID

当一个客户首次访问 Web 服务目录中的一个 JSP 页面时，Tomcat 服务器产生一个 secssion 对象。这个 session 对象调用相应的方法可以存储客户在访问该 Web 服务目录中各页面期间提交的各种信息，如姓名、号码等信息。这个 session 对象被分配了一个 String 类型的 ID，Tomcat 服务器同时将这个 ID 发送到客户机，存放在客户机的 Cookie 中。这样，session 对象和客户机之间就建立起一一对应的关系，即每个客户机都对应着一个 session 对象（该客户机的会话），不同用户的 session 对象互不相同，具有不同的 ID。当客户机再访问连接该服务目录的其他页面时，或从该服务目录连接到其他服务器再回到该服务目录时，Tomcat 服务器不再分配给客户机的新 session 对象，而是使用完全相同的一个，直到客户机关闭浏览器或这个 session 对象达到了最大生存时间，服务器取消客户机的 session 对象，即与客户机的会话对应关系消失。当客户机重新打开浏览器再连接到该服务目录时，服务器为该客户机再创建一个新的 session 对象。

注：同一个客户机在不同的服务目录中的 session 是互不相同的。

【例 4-8】客户机在服务器的某个 Web 服务目录中有 3 个页面：first.jsp、second 和 third.jsp 之间进行连接，只要不关闭浏览器，3 个页面的 session 对象是完全相同的。其中，first.jsp 存放在 Web 服务目录 chapter4 中，second.jsp 存放在目录 chapter4\tom 中，third.jsp 存放在目录 chapter4\jerry 中。客户机先访问 first.jsp 页面，从这个页面再连接到 second.jsp 页面，然后从 second.jsp 再连接到 third.jsp 页面（效果如图 4.4 所示）。

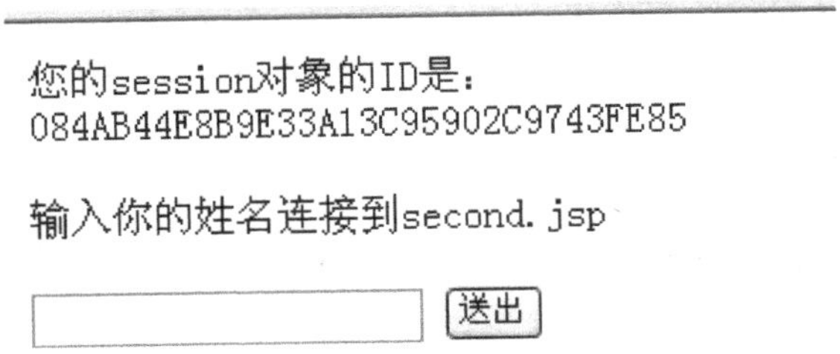

您的session对象的ID是：
084AB44E8B9E33A13C95902C9743FE85

输入你的姓名连接到second.jsp

[　　　　　] [送出]

（a）first 页面中 session 对象的 ID

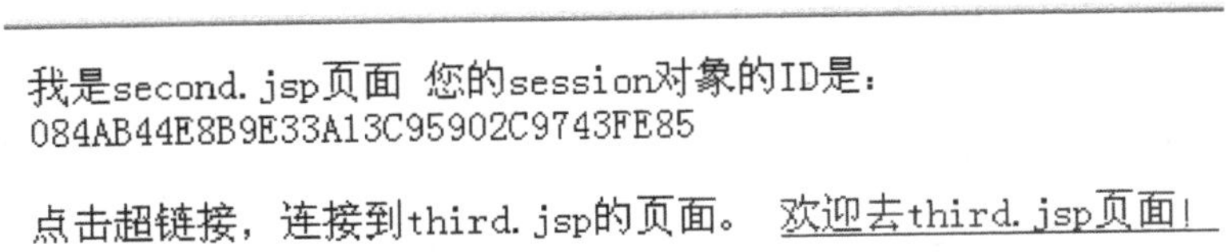

我是second.jsp页面 您的session对象的ID是：
084AB44E8B9E33A13C95902C9743FE85

点击超链接，连接到third.jsp的页面。 欢迎去third.jsp页面！

（b）second 页面中 session 对象的 ID

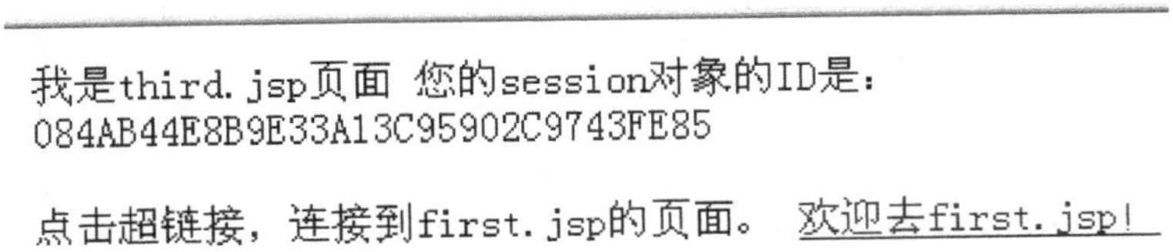

我是third.jsp页面 您的session对象的ID是：
084AB44E8B9E33A13C95902C9743FE85

点击超链接，连接到first.jsp的页面。 欢迎去first.jsp！

（c）third 页面中 session 对象的 ID

图 4.4 例 4-8 的图

first.jsp

```jsp
<%@ page contentType="text/html;charset=GB2312" %>
<HTML><BODY>
<P><% String id=session.getId();
     out.println("您的 session 对象的 ID 是： <br>"+id);
%>
<P>输入你的姓名连接到 second.jsp
   <FORM action="tom/second.jsp" method=post name=form>
       <InpuT type="text" name="boy">
       <Input type="submit" value="送出" name=submit>
   </FORM>
</BODY>
</HTML>
```

second.jsp

```jsp
<%@ page contentType="text/html;Charset=GB2312" %>
<HTML><BODY><P>我是 second.jsp 页面
```

```
        <%  String id=session.getId();
              out.println("您的 session 对象的 ID 是：<br>"+id);
        %>
    <P> 点击超链接，连接到 third.jsp 的页面。
    <A href="/chaper4/jerry/third.jsp"> 欢迎去 third.jsp 页面！</A>
    </BODY></HTML>
```

third.jsp

```
    <%@ page contentType="text/html;charset=GB2312" %>
    <HTML><BODY><P>我是 third.jsp 页面
        <%  String id=session.getId();
              out.println("您的 session 对象的 ID 是：<br>"+id);
        %>
    <P> 点击超链接，连接到 first.jsp 的页面。
    <A href="/chaper4/first.jsp">  欢迎去 first.jsp！</A>
    </BODY></HTML>
```

2．session 对象存储数据

session 对象驻留在服务器端，该对象调用某些方法保存客户机在访问某个 Web 服务目录期间的有关数据。session 对象使用下列方法处理数据。

（1）public void setAttribute(String key, Object obj)

session 对象可以调用该方法将参数 Object 指定的对象 obj 添加到 session 对象中，并为添加的对象指定了一个索引关键字，如果添加的两个对象的关键字相同，则先前添加的对象被清除。

（2）public Object getAttibute(String key)

获取 session 对象含有的关键字是 key 的对象。由于任何对象都可以添加到 session 对象中，因此用该方法取回对象时，应强制转化为原来的类型。

（3）public Enumeration getAttributeNames()

session 对象调用该方法产生一个枚举对象，该枚举对象使用 nextElemets()遍历 session 中的各对象所对应的关键字。

（4）public void removeAttribute(String name)

session 对象调用该方法移掉关键字 key 对应的对象。

当 session 对象处理数据时，非常像一个购物车。一个用户访问一个商场（类似一个 Web 服务目录）期间，分配给该用户的购物车可以放入和移出商品，用户可能在这个商场许多柜台购物（请求该 web 服务目录的各页面），当用户离开商场时（关闭浏览器，离开服务器），购物车取消。

【例 4-9】 用 session 对象模拟购物车、存储客户的姓名和购买的商品，3 个 JSP 页面都保存在 Web 服务目录 chapter4 中（效果如图 4.5 所示）。

main.jsp

```
        <%@ page contentType="text/html;Charset=GB2312" %>
        <HTML><BODY bgcolor=yellow><FONT size=2>
         <P>欢迎来到本页面，请输入您的姓名
```

（a）输入姓名

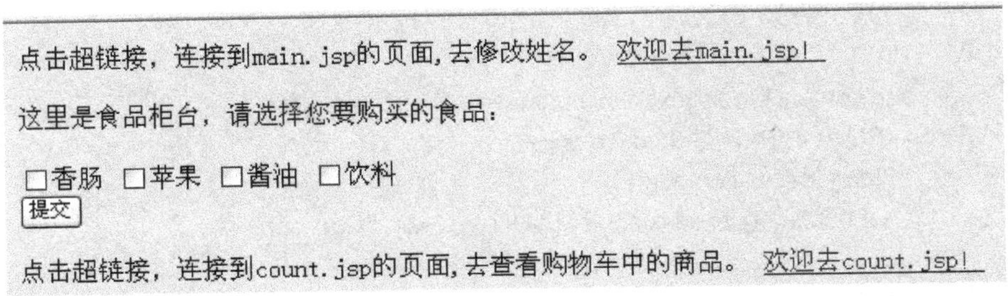

（b）选择商品

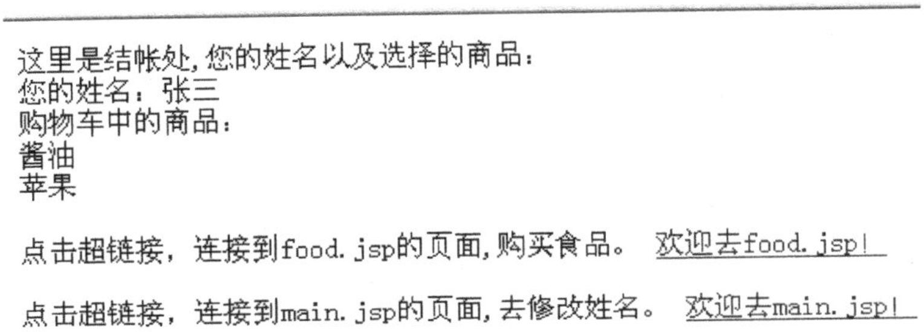

（c）查看购物车

图 4.5 例 4-9 效果图

```
<FORM action="" method=post name=form>
   <Input type="text" name="name">
   <Input type="submit" value="送出" name=submit>
</FORM>
<% String name=request.getParameter("name");
   if(name==null) {
       name="";
   }
   else{
       session.setAttribute("customerName",name);
   }
%>
<% if(name.length()>0) {
%>   <P> 点击超链接，链接到 food.jsp 的页面,去采购食品。
       <A href="food.jsp">   欢迎去食品柜台！</A>
<% }
%>
<FONT></BODY></HTML>
```

food.jsp
```
<%@ page contentType="text/html;charset=GB2312" %>
<HTML><BODY bgcolor=cyan><FONT size=3>
```

```
        <P>点击超链接，链接到 main.jsp 的页面,去修改姓名。
            <A href="main.jsp">  欢迎去 main.jsp! </A>
        <P>这里是食品柜台，请选择您要购买的食品：
          <FORM action="" method=post name=form>
            <Input type="checkbox" name="choice" value="香肠" >香肠
            <Input type="checkbox" name="choice" value="苹果" >苹果
            <Input type="checkbox" name="choice" value="酱油" >酱油
            <Input type="checkbox" name="choice" value="饮料" >饮料
            </BR>
            <Input type="submit" value="提交" name="submit">
          </FORM>
        </FONT>
        <%   String foodName[]=request.getParameterValues("choice");
             if(foodName!=null) {
                for(int k=0;k<foodName.length;k++){
                    session.setAttribute(foodName[k],foodName[k]);
                }
             }
        %>
        <P>点击超链接，链接到 count.jsp 的页面,去查看购物车中的商品。
            <A href="count.jsp">   欢迎去 count.jsp! </A>
        </BODY></HTML>
```

count.jsp
```
        <%@ page contentType="text/html;Charset=GB2312" %>
        <%@ page import="java.util.*" %>
        <HTML><P>这里是结账处,您的姓名以及选择的商品是：
        <%   String personName=(String)session.getAttribute("customerName");
             out.println("<br>您的姓名： "+personName);
             Enumeration enumGoods=session.getAttributeNames();
             out.println("<br>购物车中的商品: <br>");
             while(enumGoods.hasMoreElements()) {
                String key=(String)enumGoods.nextElement();
                String goods=(String)session.getAttribute(key);
                if(!(goods.equals(personName)))
                    out.println(goods+"<br>");
             }
        %>
        <P>点击超链接，链接到 food.jsp 的页面，购买食品。
            <A href="food.jsp">  欢迎去 food.jsp! </A>
        <P>点击超链接，链接到 main.jsp 的页面,去修改姓名。
            <A href="main.jsp">  欢迎去 main.jsp! </A>
        </FONT></BODY></HTML>
```

3．session 对象的生存期限

用户在某个 Web 服务目录的 session 对象的生存期限依赖于是否关闭浏览器，依赖于 session 对象是否调用 invalidate()方法，使得 session 无效或 session 对象达到了设置的最长的

"发呆"状态时间。如果关闭浏览器,那么用户的 session 消失。如果用户长时间不关闭浏览器,用户的 session 也可能消失,这是因为 Tomcat 服务器允许用户最长的"发呆"状态时间为 30 分钟。所谓"发呆"状态时间,是指用户对某个的 Web 服务目录发出的两次请求之间的间隔时间。比如,用户对某个 Web 服务目录下的 JSP 页面发出请求,并得到响应,如果用户不再对该 Web 服务目录发出请求(可能不再操纵浏览器或去请求其他 Web 服务目录),那么用户对该 Web 服务目录进入"发呆"状态,直到用户再次请求该 Web 服务目录时,"发呆"状态结束。

可以修改 Tomcat 服务器下的 web.xml,重新设置各 Web 服务目录下的 session 对象的最长"发呆"时间。打开 Tomcat 安装目录中 conf 文件下的配置文件 web.xml,找到

    <session-config>
      <session-timeout>30</session-timeout>
    </session-config>

将其中的"30"修改成所要求的值即可(单位为分钟)。

session 对象可以使用下列方法获取或设置生存时间有关的信息。

- public long getCreationTime():可以获取该对象创建的时间,单位是毫秒(从 1970 年 7 月 1 日午夜起至该对象创建时刻所走过的毫秒数)。
- public long getLastAccessedTime():获取当前 session 对象最后一次被操作的时间,单位是毫秒。
- public int getMaxInactiveInterval():获取 session 对象最长的"发呆"时间(单位是秒)。
- public void setMaxInactiveInterval(int interval):设置 session 对象最长的"发呆"时间(单位是秒)。
- public boolean isNew():判断当前 session 是否是一个新建的会话。
- invalidate():使 session 无效。

【例 4-10】session 对象使用 setMaxInactiveInterval(int interval)方法设置最长的"发呆"状态时间为 20 秒。用户可以通过刷新页面检查是否达到了最长的"发呆"时间,如果两次刷新之间的间隔超过 20 秒,先前的 session 将被取消,将获得一个新的 session 对象。

```
example4_10.jsp
    <%@ page contentType="text/html;Charset=GB2312" %>
    <%@ page import="java.util.*" %>
    <HTML><BODY bgcolor=yellow><FONT size=3>
     <% session.setMaxInactiveInterval(20);
        boolean boo=session.isNew();
        out.println("<br>如果你第一次访问当前 Web 服务目录,您的会话是新的");
        out.println("<br>如果你不是首次访问当前 Web 服务目录,您的会话不是新的");
        out.println("<br>会话是新的吗? :"+boo);
        out.println("<br>欢迎来到本页面,您的 session 允许的最
                         长发呆时间为"+session.getMaxInactiveInterval()+"秒");
        out.println("<br>您的 session 的创建时间是"+new Date(session.getCreationTime()));
        out.println("<br>您的 session 的 Id 是"+session.getId());
        Long lastTime=(Long)session.getAttribute("lastTime");
        if(lastTime==null) {
            long n=session.getLastAccessedTime();
```

```
                session.setAttribute("lastTime",new Long(n));
            }
            else{
                long m=session.getLastAccessedTime();
                long n=((Long)session.getAttribute("lastTime")).longValue();
                out.println("<br>您的发呆时间大约是"+(m-n)+"毫秒，大约"+(m-n)/1000+"秒");
                session.setAttribute("lastTime",new Long(m));
            }
        %>
    <FONT></BODY></HTML>
```

4. session 对象与 URL 重写

session 对象能与客户机建立起一一对应关系依赖于用户的浏览器是否支持 Cookie。如果客户机不支持 Cookie，那么不同网页之间的 session 对象可能是互不相同的，因为服务器无法将 ID 存放到客户机，就不能建立 session 对象与客户机的一一对应关系。我们将浏览器的 Cookie 设置为禁止后（选择"浏览器菜单"→"工具"→"Internet 选项"→"隐私"，将第三方 Cookie 设置成禁止），再运行前面的例 4-8，会得到不同的结果。即"同一客户机"对应了多个 session 对象，这样服务器就无法知道在这些页面上访问的实际上是同一个客户机。

如果客户机的浏览器不支持 Cookie，可以通过 URL 重写来实现 session 对象的唯一性。所谓 URL 重写，就是当客户机从一个页面重新连接到一个页面时，通过向这个新的 URL 添加参数，把 session 对象的 ID 传带过去，这样就可以保障客户机在该网站各页面中的 session 对象是完全相同的。可以使用 response 对象调用 encodeURL()或 encodeRedirectURL()方法实现 URL 重写。比如，如果从 tom.jsp 页面连接到 jerry 页面，首先实现 URL 重写：

      String str=response.encodeRedirectURL("jerry.jsp");

然后将连接目标写成"<%=str%>"。

【例 4-11】 如果客户机不支持 Cookie，可以将例 4-8 中的 first.jsp、second.jsp 和 third.jsp 实行 URL 重写。

first.jsp
```
    <%@ page contentType="text/html;charset=GB2312" %>
    <HTML><BODY>
      <P><%   String id=session.getId();
              out.println("您的 session 对象的 ID 是：<br>"+id);
              String str=response.encodeRedirectURL("tom/second.jsp");
          %>
      <P>输入你的姓名连接到 second.jsp
        <FORM action="<%=str%>" method=post name=form>
          <Input type="text" name="boy">
          <Input type="submit" value="送出" name=submit>
        </FORM>
    </BODY></HTML>
```

second.jsp
```
    <%@ page contentType="text/html;Charset=GB2312" %>
    <HTML><BODY>
```

```
            <P>我是 second.jsp 页面
                <%    String id=session.getId();
                    out.println("您的 session 对象的 ID 是: <br>"+id);
                    String str=response.encodeRedirectURL("/chaper4/jerry/third.jsp");
                %>
            <P> 点击超链接,连接到 third.jsp 的页面。
        <A href="<%=str%>"> 欢迎去 third.jsp 页面! </A>
        </BODY></HTML>
```

third.jsp

```
        <%@ page contentType="text/html;charset=GB2312" %>
        <HTML><BODY>
          <P>我是 third.jsp 页面
            <% String id=session.getId();
                out.println("您的 session 对象的 ID 是: <br>"+id);
                String str=response.encodeRedirectURL("/chaper4/first.jsp");
            %>
          <P> 点击超链接,连接到 first.jsp 的页面。
        <A href="<%=str%>">   欢迎去 first.jsp! </A>
        </BODY></HTML>
```

5. 计数器

计数器可以记录某个 Web 服务目录(通常所说的网站)被不同用户的浏览器访问次数,但需要限制客户通过不断刷新页面或再次访问其他的页面来增加计数器的计数。当一个用户请求该 Web 服务目录下的任何一个 JSP 页面时,首先检查用户的 session 对象中是否已经有计数,如果没有计数,立刻将当前的计数增 1,并存到客户的 session 中。

【例 4-12】 Web 服务目录有 2 个 JSP 页面 helloOne.jsp、helloTwo.jsp 和 1 个 tag 文件 count.tag。count.tag 文件负责计数。helloOne.jsp、helloTwo.jsp 使用 count.tag 实现计数。用户首次请求 helloOne.jsp 和 helloTwo.jsp 的任何一个,都会使得网站的计数增 1。

helloOne.jsp

```
        <%@ page contentType="text/html;Charset=GB2312" %>
        <%@ taglib prefix="person" tagdir="/WEB-INF/tags" %>
        <HTML><BODY>
          <P>Welcome 欢迎您访问本站
            <person:count/>
            <A href="helloTwo.jsp">欢迎去 helloTwo.jsp 参观</A>
        </BODY></HTML>
```

helloTwo.jsp

```
        <%@ page contentType="text/html;Charset=GB2312" %>
        <%@ taglib prefix="person" tagdir="/WEB-INF/tags" %>
        <HTML><BODY>
          <P>Welcome 欢迎您访问本站
            <person:count/>
            <A href="helloTwo.jsp">欢迎去 helloOne.jsp 参观</A>
        </BODY>
        </HTML>
```

count.tag

```jsp
<%@ tag import="java.io.*" %>
<FONT size=4>
<%!  int number=0;
     File file=new File("count.txt") ;
     synchronized void countPeople(){                //计算访问次数的同步方法
         if(!file.exists()) {
             number++;
             try {
                 file.createNewFile();
                 FileOutputStream out=new FileOutputStream("count.txt");
                 DataOutputStream dataOut=new DataOutputStream(out);
                 dataOut.writeInt(number);
                 out.close();
                 dataOut.close();
             }
             catch(IOException ee) { }
         }
         else{
             try{
                 FileInputStream in=new FileInputStream("count.txt");
                 DataInputStream dataIn=new DataInputStream(in);
                 number=dataIn.readInt();
                 number++;
                 in.close();
                 dataIn.close();
                 FileOutputStream out=new FileOutputStream("count.txt");
                 DataOutputStream dataOut=new DataOutputStream(out);
                 dataOut.writeInt(number);
                 out.close();
                 dataOut.close();
             }
             catch(IOException ee) { }
         }
     }
%>
<%   String str=(String)session.getAttribute("count");
     if(str==null) {
         countPeople();
         String personCount=String.valueOf(number);
         session.setAttribute("count",personCount);
     }
%>
<P><P>您是第
       <%=(String)session.getAttribute("count")%>
   个访问本网站的客户。
</FONT>
```

4.4 out 对象

out 对象是一个输出流,指向客户的浏览器的缓存区,out 对象调用相应的方法可以将数据发送到客户端浏览器的缓存中。内置对象 out 对象是 JspWriterout 类的一个实例,可以在 Toamcat 服务器的 webapps\tomcat-docs\servletapi 中查找 JspWriterout 类的方法。

out 对象可调用如下方法用于各种数据的输出,例如:

out.print(Boolean),out.println(boolean)	用于输出一个布尔值
out.print(char),out.println(char)	输出一个字符
out.print(double),out.println(double)	输出一个双精度的浮点数
out.print(fload),out.println(float)	用于输出一个单精度的浮点数
out.print(long),out.println(long)	输出一个长整型数据
out.print(String),out.println(String)	输出一个字符串对象的内容
out.flush()	输出缓冲区里的内容
out.close()	关闭流

方法 println()和 print()的区别是:println()会向缓存区写入一个换行,而 print()不写入换行。但是浏览器的显示区域目前不识别 println()写入的换行,如果希望浏览器显示换行,应当向浏览器写入
实现换行。

【例 4-13】 使用 out 对象向客户输出包括表格等内容的信息。

example4_13.jsp

```
<%@ page contentType="text/html;charset=GB2312" %>
<%@ page import="java.util.*" %>
<HTML><BODY>
  <% int a=2200;long b=3456;boolean c=true;
     out.println(a);
     out.println(b);
     out.print("<br>");
     out.println(c);
  %>
<Left>   <P><FONT size=2 >以下是一个表格</FONT>
<%   out.print("<Font face=隶书  size=2 >");
     out.println("<table Border=1>");
     out.println("<tr >");
     out.println("<th width=80>"+"姓名"+"</th>");
     out.println("<th width=60>"+"性别"+"</th>");
     out.println("<th width=200>"+"出生日期"+"</th>");
     out.println("</tr>");
     out.println("<tr >");
     out.println("<td >"+"张三"+"</td>");
     out.println("<td >"+"男"+"</td>");
     out.println("<td >"+"1988 年 5 月"+"</td>");
     out.println("</tr>");
     out.println("<tr >");
```

```
            out.println("<td >"+"李四"+"</td>");
            out.println("<td >"+"男"+"</td>");
            out.println("<td >"+"1987 年 10 月"+"</td>");
            out.println("</tr>");
            out.println("</Table>");
            out.print("</Font>")  ;
        %>
    </Center></BODY></HTML>
```

4.5 application 对象

我们已经知道，不同用户的 session 对象互不相同。但有时候，用户之间可能需要共享一个对象，Tomcat 服务器启动后，就产生了这样一个对象，它就是内置对象 application。任何用户在所访问的服务目录的各页面时，这个 application 对象都是同一个，直到服务器关闭，这个 application 对象才被取消。

1. application 对象的常用方法

（1）public void setAttribute(String key, Object obj)

application 对象可以调用该方法将参数 Object 指定的对象 obj 添加到 application 对象中，并为添加的对象指定了一个索引关键字，如果添加的两个对象的关键字相同，则先前添加对象被清除。

（2）public Object getAttibue(String key)

获取 application 对象含有的关键字是 key 的对象。由于任何对象都可以添加到 application 对象中，因此用该方法取回对象时，应强制转化为原来的类型。

（3）public Enumeration getAttributeNames()

application 对象调用该方法产生一个枚举对象，该枚举对象使用 nextElemets()遍历 application 中的各对象所对应的关键字。

（4）public void removeAttribue(String key)

从当前 application 对象中删除关键字是 key 的对象。

（5）public String getServletInfo()

获取 Servlet 编译器的当前版本的信息。

由于 application 对象对所有的客户都是相同的，任何客户机对该对象中存储的数据的改变都会影响到其他客户机，因此在某些情况下，对该对象的操作需要实现同步处理。

注：有些服务器不直接支持使用 application 对象，必须用 ServletContext 类声明这个对象，再使用 getServletContext()方法对这个 application 对象进行初始化。

2. 用 application 制作留言板

在下面的例 4-14 中，客户机通过 submit.jsp 向 messagePane.jsp 页面提交姓名、留言标题和留言内容，messagePane.jsp 页面获取这些内容后，用同步方法将这些内容添加到一个向量中，然后将这个向量添加到 application 对象中。当用户查看留言板时，showMessage.jsp 负责显示所有用户的留言内容，即从 application 对象中取出向量，然后遍历向量中存储的信息。

在这里使用了向量这种数据结构，Java 的 java.util 包中的 Vector 类负责创建一个向量对象。如果你已经学会使用数组，那么很容易就会使用向量。创建一个向量时不用像数组那样必须要给出数组的大小。向量创建后，如"Vector a=new Vector()"，a 可以使用 add(Object o) 方法把任何对象添加到向量的末尾，向量的大小会自动增加。可以使用 add(int index, Object o) 方法把一个对象追加到该向量的指定位置。向量 a 可以使用 elementAt(int index)方法获取指定索引处的向量的元素（索引初始位置是 0）；a 可以使用 size()方法获取向量所含有的元素的个数。另外，与数组不同的是向量的元素类型不要求一致。需要注意的是，虽然可以把任何一种 Java 的对象放入一个向量，但是当从向量中取出一个元素时，必须使用强制类型转化运算符将其转化为原来的类型。

【例 4-14】 （效果如图 4.6 所示）

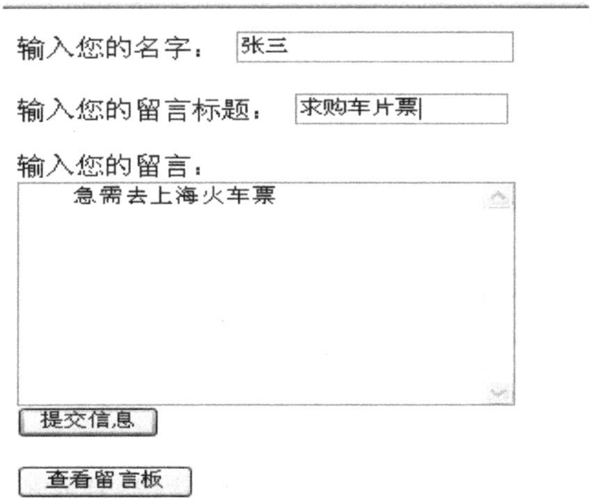

（a）输入留言信息

（b）信息存储到留言板

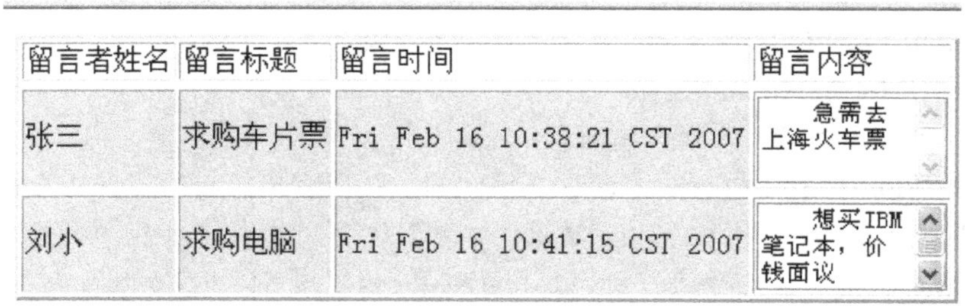

（c）查看留言

图 4.6 例 4-14 效果图

submit.jsp

```
<%@ page contentType="text/html;charset=GB2312" %>
<HTML><BODY>
  <FORM action="messagePane.jsp" method="post" name="form">
   <P>输入您的名字：
    <Input  type="text" name="peopleName">
```

```
            <BR>输入您的留言标题：
        <Input   type="text"   name="Title">
            <BR>输入您的留言：
            <BR> <TextArea name="messages" ROWs="10" COLS=36 WRAP="physical" >
        </TextArea>
            <BR> <Input type="submit" value="提交信息" name="submit">
    </FORM>
    <FORM action="showMessage.jsp" method="post" name="form1">
        <Input type="submit" value="查看留言板" name="look">
    </FORM>
</BODY></HTML>
```

messagePane.jsp

```
<%@ page contentType="text/html;Charset=GB2312" %>
<%@ page import="java.util.*" %>
<HTML><BODY>
    <%! Vector v=new Vector();
        ServletContext application;
        synchronized void sendMessage(String s) {
            application=getServletContext();;
            v.add(s);
            application.setAttribute("Mess",v);
        }
    %>
    <% String name=request.getParameter("peopleName");
        String title=request.getParameter("Title");
        String messages=request.getParameter("messages");
        if(name==null) {
            name="guest"+(int)(Math.random()*10000);
        }
        if(title==null) {
            title="无标题";
        }
        if(messages==null) {
            messages="无信息";
        }
        String time=new Date().toString();
        String s="#"+name+"#"+title+"#"+time+"#"+messages+"#";
        sendMessage(s);
        out.print("您的信息已经提交！ ");
    %>
    <A href="submit.jsp" >返回
    <A href="showMessage.jsp" >查看留言板
</BODY></HTML>
```

showMessage.jsp

```
<%@ page contentType="text/html;Charset=GB2312" %>
<%@ page import="java.util.*" %>
```

```
<HTML><BODY>
   <% Vector v=(Vector)application.getAttribute("Mess");
       out.print("<table border=2>");
       out.print("<tr>");
       out.print("<td bagcolor=cyan>"+"留言者姓名"+"</td>");
       out.print("<td bagcolor=cyan>"+"留言标题"+"</td>");
       out.print("<td bagcolor=cyan>"+"留言时间"+"</td>");
       out.print("<td bagcolor=cyan>"+"留言内容"+"</td>");
       out.print("</tr>");
       for(int i=0;i<v.size();i++) {
           out.print("<tr>");
           String message=(String)v.elementAt(i);
           StringTokenizer fenxi=new StringTokenizer(message,"#");
           out.print("<tr>");
           int number=fenxi.countTokens();
           for(int k=0;k<number;k++) {
               String str=fenxi.nextToken();
               if(k<number-1) {
                   out.print("<td bgcolor=cyan >"+str+"</td>");
               }
               else{
                   out.print("<td><TextArea rows=3 cols=12>"+str+"</TextArea></td>");
               }
           }
       out.print("</tr>");
       out.print("</table>");
   %>
</BODY></HTML>
```

习 题 4

1. request 对象经常使用 getParameter(String s)方法获取请求中的哪部分信息？
2. 如果表单提交的信息中有中文，接收该信息的页面应做怎样的处理？
3. 编写 2 个 JSP 页面 inputNumber.jsp 和 computeAndShow.jsp，用户可以使用 inputNumber.jsp 提供的表单输入一个数字，并提供给 computeAndShow.jsp 页面。该页面通过内置对象获取 inputNumber.jsp 页面中提交的数字，计算并显示该数的平方和立方值。
4. response 调用 sendRedirect(URL url)方法的作用是什么？
5. 在第 2 题的基础上，增加如下功能：如果用户在 inputNumber.jsp 页面中输入了非数字，那么 computeAndShow.jsp 页面提示用户输入错误，将用户重新定向到 inputNumber.jsp 页面。
6. 用户在不同 Web 服务目录中的 session 对象是相同的吗？用户在同一 Web 服务目录的不同子目录中的 session 对象是相同的吗？
7. 改进例 4-9，在 main.jsp 页面中，用户可以输入年龄和性别。
8. 改进例 4-14，将留言板的留言时间用 yyyy-mm-dd hh:mm:ss 的形式来显示。

第 5 章

JSP 与 JavaBean

> **本章导读**
> ✿ 知识点：掌握怎样编写 JavaBean 和使用 JavaBean。理解怎样使用 JavaBean 分离 JSP 页面的数据显示和数据处理。
> ✿ 重点：理解各种 JavaBean 的生命周期。使用有关标记加载 JavaBean、设置和获取 JavaBean 的属性的值
> ✿ 难点：掌握 session 声明周期的 JavaBean 的使用。
> ✿ 关键实践：针对某实际问题编写 JSP 页面，使用 JavaBean 处理有关数据，将数据的显示和处理分离。

　　JSP 页面中的 HTML 标记是页面的静态部分，即不需要服务器做任何处理，直接发送给客户的信息，JSP 页面通过使用 HTML 标记，可以为用户提供一个友好的界面，即所谓的数据表示层。而 JSP 页面中的变量声明、程序片以及表达式为动态部分，需要服务器做出处理后，再将有关处理后的结果发送给客户。编写一个健壮的 Web 应用程序，提倡将数据的表示和处理分离，如果一个 JSP 页面将数据表示和处理混杂在一起将导致代码混乱，不利于 Web 应用的扩展和维护。在第 3 章，我们曾学习了 Tag 文件，Tag 文件是可复用代码，可供不同的 JSP 页面使用，JSP 页面通过使用 Tag 文件来实现数据的表示和处理分离。本章介绍 JavaBean，JSP 页面通过 JavaBean 也可以实现数据的表示和处理的分离。

　　按照 Sun 公司的定义，JavaBean 是一个可重复使用的软件组件，是遵循一定标准、用 Java 语言编写的一个类，该类的一个实例称为一个 JavaBean，简称 bean。由于 JavaBean 是基于 Java 语言的，因此 JavaBean 不依赖平台，具有以下特点：

- ⊙ 可以实现代码复用。
- ⊙ 易编写、易维护、易使用。
- ⊙ 可以在任何安装了 Java 运行环境的平台上的使用，而不需要重新编译。

　　一个 JSP 页面可以将数据的处理过程指派给一个或几个 bean 来完成，只需在 JSP 页面中调用这个 bean 即可。在 JSP 页面中调用 bean，可以将数据的处理代码从页面中分离出来，

实现代码复用,更有效地维护一个 Web 应用。

本章在 webapps 目录下新建一个 Web 服务目录 chapter5。因此,除非特别约定,本章例子中涉及的 JSP 页面均保存在 chapter5 中。

5.1 编写和使用 JavaBean

5.1.1 编写 bean

JavaBean 分为可视组件和非可视组件。JSP 中主要使用非可视组件。对于非可视组件,我们不必去设计它的外观,主要关心它的属性和方法。

编写 JavaBean 就是编写一个 Java 的类,这个类创建的一个对象称为一个 bean。为了能让使用这个 bean 的应用程序构建工具(如 Tomcat 服务器)知道这个 bean 的属性和方法,只需在类的方法命名上遵守以下规则。

① 如果类的成员变量的名字是 xxx,那么为了获取或更改成员变量的值,即获取或更改属性,类中必须提供两个方法:

    getXxx()   //用来获取属性 xxx
    setXxx()   //用来修改属性 xxx

即方法的名字用 get 或 set 为前缀,后缀是将成员变量名字的首字母大写的字符序列。

② 对于 boolean 类型的成员变量,即布尔逻辑类型的属性,允许使用"is"代替上面的"get"和"set"。

③ 类中声明的方法的访问属性都必须是 public 的。

④ 类中声明的构造方法必须是 public、无参数的。

下面编写一个简单的 bean,并说明在 JSP 中怎样使用这个 bean。如果使用 Tomcat 5.x 服务器,bean 必须带有包名。使用 package 语句给 bean 一个包名。包名可以是一个合法的标识符,也可以是若干个标识符加"."分割而成,例如:

    package gping;
    package tom.jiafei;

以下是一个可以用来创建 bean 的类,该类创建的 bean 可以计算梯形的面积。

创建 bean 的源文件如下:

Lader.java

```
    package tom.jiafei;
    public class Lader {
        double above,bottom,area;
        public Lader() {
            above=1;
            bottom=1;
            area=(above+bottom)/2.0;
        }
        public double getAbove() {
            return above;
        }
        public void setAbove(double above) {
```

```
            this.above=above;
        }
        public double getBottom (){
            return bottom;
        }
        public void setBottom(double bottom) {
            this.bottom=bottom;
        }
        public double getArea() {
            area=(above+bottom)/2.0;
            return area;
        }
    }
```

注：Lader 类中的 area 属性是一个关联属性，即 area 的值依赖于 above 和 bottom，所以不应该提供 setArea()方法，只需提供 getArea()方法即可。

将上述 Java 文件保存为 Lader.java，并编译通过，得到字节码文件 Lader.class。

5.1.2 使用 bean

为了使 JSP 页面使用 bean，Tomcat 服务器必须使用相应的字节码创建一个对象，即创建一个 bean。为了让 Tomcat 服务器能找到字节码，字节码文件必须存放在特定的目录中。

1. 字节码文件的保存目录

首先，在当前 Web 服务目录下建立如下目录结构：

   Web 服务目录\WEB-INF\classes

然后根据类的包名，在目录 classes 下建立相应的子目录，如类的包名为 tom.jiafei，那么在 classes 下建立子目录 tom\jiafei。为了让 Tomcat 服务器启用上述目录，必须重新启动 Tomcat 服务器。Tomcat 服务器重新启动后，就可以将创建 bean 的字节码文件，如 Lader.class，复制到"Web 服务目录\WEB-INF\classes\tom\jiafei"中。

2. 在 JSP 页面中使用 bean

为了在 JSP 页面中使用 bean，必须使用 JSP 动作标记 useBean。useBean 标记的格式如下：

   <jsp:useBean id="给 bean 起的名字" class="创建 bean 的类" scope="bean 有效范围">
   </jsp:useBean>

或

   <jsp:useBean id="给 bean 起的名字" class="创建 bean 的类" scope="bean 有效范围"/>

当服务器上某个含有 useBean 动作标记的 JSP 页面被加载执行时，Tomcat 服务器将首先根据 id 的名字，在一个同步块中，查找 Tomcat 服务器内置的 pageContent 对象中是否含有名字 id 和作用域 scope 的对象，如果这个对象存在，Tomcat 服务器就这个对象的一个副本（即 bean）分配给 JSP 页面。这样，JSP 页面就获得了一个作用域是 scope、名字是 id 的 bean（就像我们组装电视机时获得了一个有一定功能和使用范围的电子元件）。如果在 pageContent 中没有查找到指定作用域、名字是 id 的对象，就根据 class 指定的类创建一个名字是 id 对象，即创建了一个名字是 id 的 bean，并添加到 pageContent 内置对象中，指定该 bean 的作用域是

scope，同时 Tomcat 服务器分配给 JSP 页面一个作用域是 scope、名字是 id 的 bean。

从 Tomcat 服务器创建 bean 的过程可以看出，首次创建一个新的 bean 需要用相应的字节码文件创建对象，当某些 JSP 页面再需要同样的 bean 时，Tomcat 服务器直接将内置的 pageContent 中已经有的对象的副本分配给 JSP 页面，提高了 JSP 使用 bean 的效率。

注：如果修改了字节码文件，必须重新启动 Tomcat 服务器，才能使用新的字节码文件。

下面就 useBean 标签中 scope 取值的不同情况阐述如下。

（1）scope 取值 page

Tomcat 服务器分配给每个 JSP 页面的 bean 是互不相同的，也就是说，尽管每个 JSP 页面的 bean 的功能相同，但占有的内存空间不同。该 bean 的有效范围是当前页面，当客户离开这个页面时，Tomcat 取消分配的 bean，即释放 bean 所占有的内存空间。

需要注意的是，不同用户的 scope 取值是 page 的 bean 也是互不相同的，即当两个用户同时访问一个 JSP 页面时，一个用户对自己 bean 的属性的改变，不会影响到另一个用户。

（2）scope 取值 session

bean 的有效范围是客户机的会话期间，也就是说，如果客户机在多个页面中相互连接，每个页面都含有一个 useBean 标记，而且各页面的 useBean 标签中 id 的值相同、scope 的值都是 session，那么该客户机在这些页面得到的 bean 是相同的一个（占有相同的内存空间）。如果客户机在某个页面更改了这个 bean 的属性，其他页面的这个 bean 的属性也将发生同样的变化。当客户机的会话（session）消失，如关闭浏览器时，Tomcat 取消分配的 bean，即释放 bean 所占有的内存空间。

不同用户的 scope 取值是 session 的 bean 是互不相同的（占有不同的内存空间），即两个用户同时访问一个 JSP 页面时，一个用户对自己 bean 的属性的改变不会影响到另一个用户。

（3）scope 取值 request

bean 的有效范围是 request 期间。客户机在网站的访问期间可能请求过多个页面，如果这些页面含有 scope 取值是 request 的 useBean 标记，那么在每个页面分配的 bean 也是互不相同的。Tomcat 服务器对请求做出响应之后，取消分配给 JSP 页面的这个 bean。

不同用户的 scope 取值是 request 的 bean 也是互不相同的，也就是说，当两个用户同时请求一个 JSP 页面时，一个用户对自己 bean 的属性的改变，不会影响到另一个用户。

（4）scope 取值 application

Tomcat 服务器为所有的 JSP 页面分配一个共享的 bean，不同用户的 scope 取值是 application 的 bean 也都是相同的一个，也就是说，当几个用户同时访问一个 JSP 页面时，任何一个用户对自己 bean 的属性的改变都会影响到其他用户。

注：当使用作用域是 session 的 bean 时，要保证客户机支持 Cooker。

在使用 bean 的 JSP 页面中，必须有相应的 import 指令，例如：

    <%@ page import="tom.jiafei.*"%>

【**例 5-1**】useBean.jsp 存放在 Web 服务目录 chaper5 中，负责创建 bean 的类是上述的 Lader 类，存放在 chaper5\WEB-INF\classes\tom\jiafei 中。创建的 bean 的名字是 lader，lader 的 scope 取值是 page。（效果如图 5.1 所示）

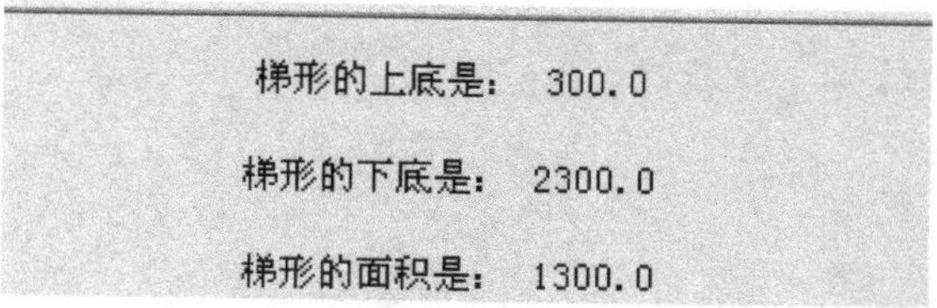

图 5.1 使用 bean 的 JSP 页面

useBean.jsp

```
<%@ page contentType="text/html;Charset=GB2312" %>
<%@ page import="tom.jiafei.*"%>
<HTML>
<BODY bgcolor=cyan><FONT size=5>
  <jsp:useBean id="lader" class="tom.jiafei.Lader" scope="page" />
  <%--通过上述 JSP 标记，客户获得了一个作用域是 page，名字是 lader 的 bean --%>
  <%  //设置梯形的上、下底：
      lader.setAbove(300);
      lader.setBottom(2300);
  %>
<P>梯形的上底是：<%=lader.getAbove()%>
<P>梯形的下底是：<%=lader.getBottom()%>
<P>梯形的面积是：<%=lader.getArea()%>
</BODY></HTML>
```

在下面的例 5-2 中，我们将 bean 的 scope 的值设为 session，创建的 bean 的名字是 lader，创建该 bean 的类文件仍然是上述的 Lader.class。在 bean1.jsp 页面中，lader 的 above 和 bottom 都是默认值 1（如图 5.2 所示），然后链接到 bean2.jsp 页面，将 lader 的 above 和 bottom 值更改为 600 和 800（如图 5.3 所示）。再链接到 bean1.jsp，会发现 lader 的 above 和 bottom 的值已经改变为 600 和 800（如图 5.4 所示）。

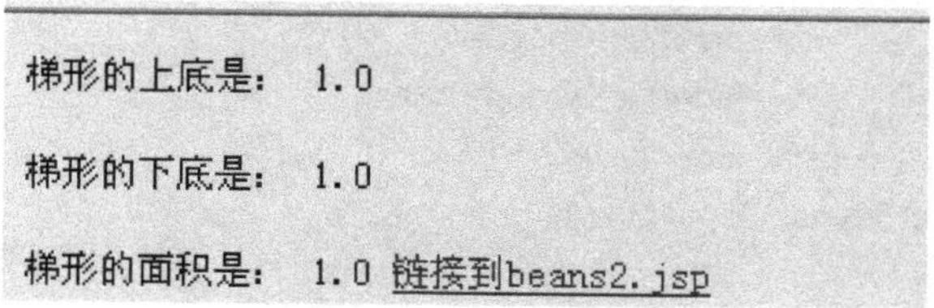

图 5.2 bean1.jsp 效果

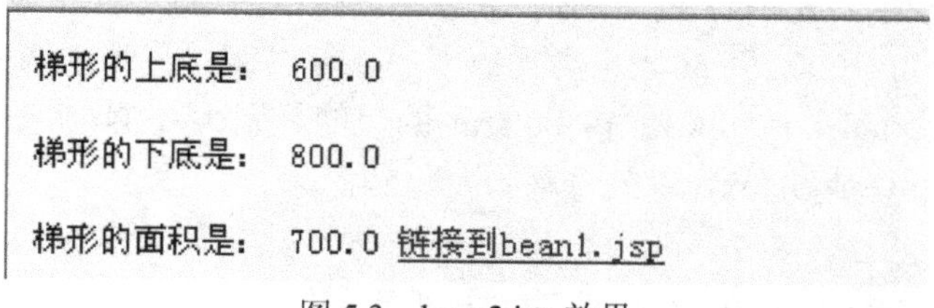

图 5.3 bean2.jsp 效果

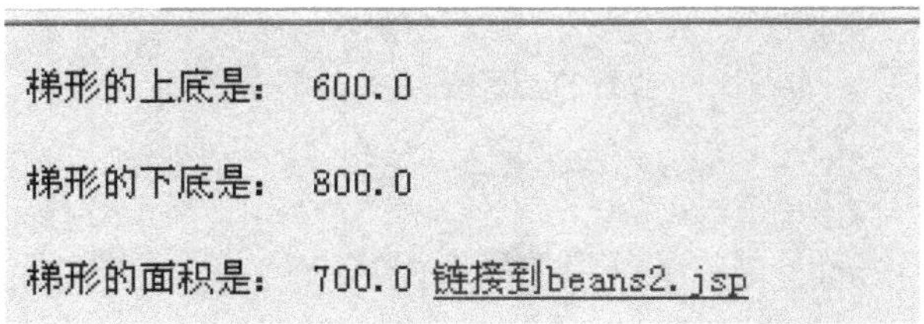

```
梯形的上底是：  600.0

梯形的下底是：  800.0

梯形的面积是：  700.0  链接到beans2.jsp
```

图 5.4 刷新 bean1.jsp 后效果

【例 5-2】 beans1.jsp 页面存放在 Web 服务目录 chapter5 中，beans2.jsp 页面存放在目录 chapter5\hello 中。

bean1.jsp

```jsp
<%@ page contentType="text/html;Charset=GB2312" %>
<%@ page import="tom.jiafei.*"%>
<HTML><BODY bgcolor=cyan><FONT size=5>
    <jsp:useBean id="lader" class="tom.jiafei.Lader" scope="session" />
<P>梯形的上底是：  <%=lader.getAbove()%>
<P>梯形的下底是：  <%=lader.getBottom()%>
<P>梯形的面积是：  <%=lader.getArea()%>
<A href="hello/bean2.jsp">链接到 beans2.jsp</A>
</BODY></HTML>
```

bean2.jsp

```jsp
<%@ page contentType="text/html;Charset=GB2312" %>
<%@ page import="tom.jiafei.*"%>
<HTML><BODY bgcolor=yellow><FONT size=5>
    <jsp:useBean id="lader" class="tom.jiafei.Lader" scope="session" />
    <% lader.setAbove(600);
       lader.setBottom(800);
    %>
<P>梯形的上底是：  <%=lader.getAbove()%>
<P>梯形的下底是：  <%=lader.getBottom()%>
<P>梯形的面积是：  <%=lader.getArea()%>
<A href="/chaper5/bean1.jsp">链接到 bean1.jsp</A>
</BODY></HTML>
```

【例 5-3】 将 bean 的 scope 的值设为 application，创建的 bean 的名字是 lader，创建该 bean 的类文件仍然是上述的 Lader.class。当第一个客户访问页面 door1.jsp 时，把 lader 的 above 和 bottom 的值修改为 1000 和 5000。当其他客户（可以新打开浏览器模拟其他用户）访问 door2.jsp 时，看到的 lader 的 above 和 bottom 的值的分别为 1000 和 5000。本例中的页面 door1.jsp 和 door2.jsp 存放在 Web 服务目录 chaper5 中。

door1.jsp

```jsp
<%@ page contentType="text/html;Charset=GB2312" %>
<%@ page import="tom.jiafei.*"%>
<HTML><BODY bgcolor=cyan><FONT size=5>
    <jsp:useBean id="lader" class="tom.jiafei.Lader" scope="application" />
```

```
        <% lader.setAbove(1000);
            lader.setBottom(5000);
        %>
        <P>梯形的上底是：<%=lader.getAbove()%>
        <P>梯形的下底是：<%=lader.getBottom()%>
        </BODY></HTML>
```

door2.jsp
```
        <%@ page contentType="text/html;Charset=GB2312" %>
        <%@ page import="tom.jiafei.*"%>
        <HTML><BODY bgcolor=yellow><FONT size=5>
          <jsp:useBean id="lader" class="tom.jiafei.Lader" scope="application" />
          <P>梯形的上底是：   <%=lader.getAbove()%>
          <P>梯形的下底是：   <%=lader.getBottom()%>
        </BODY> </HTML>
```

5.2 获取和修改 bean 的属性

当使用 useBean 动作标记创建一个 bean 后，在 Java 程序片中，这个 bean 就可以调用方法产生行为，如修改属性、使用类的中的方法等，如前面的例 5-1～例 5-3 所示。

获取或修改 bean 的属性还可以使用动作标记 getProperty、setProperty，一个 bean 就是一个遵守了一定规范的类所创建的对象，当 JSP 页面使用<jsp:useBean>标记获得一个 bean 之后，就可以使用<jsp:setProperty>和<jsp:getProperty>标记设置和获取 bean 的属性，在 JSP 页面中不必使用 Java 程序片。下面讲述怎样使用 JSP 的动作标记去获取和修改 bean 的属性。

5.2.1 动作标记 getProperty

使用 getProperty 标记可以获得 bean 的属性值，并将这个值用串的形式显示给客户。使用这个标记之前，必须使用 useBean 标记获取得到一个 bean。

getProperty 动作标记格式如下：

    <jsp:getProperty   name= "bean 的名字"   property= "bean 的属性" />

或

    <jsp:getProperty   name= "bean 的名字"   property= "bean 的属性"/>
    </jsp:getProperty>

该指令的作用相当于在程序片中使用 bean 调用 getXxx()方法。其中，name 取值是 bean 的名字，用来指定要获取哪个 bean 的属性的值；property 取值是该 bean 的一个属性的名字。

以下是一个描述图书基本信息的类 Book.java。

Book.java
```
        package tom.jiafei;
        public class Book {
            String name,ISBN;
            float price;
            public String getName() {
```

```
            return name;
        }
        public void setName(String newName) {
            name=newName;
        }
        public String getISBN() {
            return ISBN;
        }
        public void setISBN(String newISBN) {
            ISBN=newISBN;
        }
        public float getPrice() {
            return price;
        }
        public void setPrice(float newPrice) {
            price=newPrice;
        }
    }
```

【例 5-4】 book.jsp 页面保存在 Web 服务目录 chapter5 中，该页面使用 useBean 标记，用 Book 类创建一个名字是 book 的 bean，并使用 getProperty 标记获取 book 的各属性的值。效果如图 5.5 所示。

```
书名:    null
ISBN是:  null
价钱:    0.0
书名:    JSP教程
ISBN是:  7-302-08867-8
价钱:    26.9
```

图 5.5　使用 getProperty 标记获取 bean 属性值

book.jsp

```
<%@ page contentType="text/html;Charset=GB2312" %>
<%@ page import="tom.jiafei.*"%>
<HTML>
<BODY bgcolor=cyan><FONT size=4>
    <jsp:useBean id="book" class="tom.jiafei.Book" scope="request" />
<BR>书名： <jsp:getProperty  name= "book"   property= "name"   />
<BR>ISBN 是:
    <jsp:getProperty name= "book"   property= "ISBN" />
<BR>价钱：   <jsp:getProperty name= "book"   property= "price" />
<%   book.setISBN("7-302-08867-8");
     book.setName("JSP 教程");
     book.setPrice(26.9f);
```

```
%>
<BR>书名：   <jsp:getProperty  name="book"   property="name" />
<BR>ISBN 是： <jsp:getProperty name="book"   property="ISBN" />
<BR>价钱：    <jsp:getProperty name="book"   property="price" />
</BODY> </HTML>
```

5.2.2 动作标记 setProperty

使用 setProperty 标记可以设置 bean 的属性值。使用这个标记之前，必须使用 useBean 标记得到一个可操作的 bean。

setProperty 动作标记可以通过 3 种方式设置 bean 属性的值。

1．将 beans 属性的值设置为一个表达式的值或字符串

使用 setProperty 动作标记将 bean 的属性设置为一个表达式的值的格式如下：

   `<jsp:setProperty   name="bean 的名字" property="bean 的属性" value="<%=expression%>" />`

使用 setProperty 动作标记将 bean 的属性设置为一个字符串的格式如下：

   `<jsp:setProperty   name="bean 的名字" property="bean 的属性" value=字符串 />`

如果将表达式的值设置为 bean 属性的值，表达式值的类型必须与 bean 的属性的类型一致。如果将字符串设置为 bean 属性的值，这个字符串会自动被转化为 bean 的属性的类型。Java 语言将字符串转化为其他数值类型的方法如下：

- 转化到 int：Integer.parseInt(Sting s)。
- 转化到 long：Long.parseLong(Sting s)。
- 转化到 float：Float.parseFloat(Sting s)。
- 转化到 double：Double.parseDouble(Sting s)。

这些方法都可能发生 NumberFormatException 异常，如将字符串"ab23"转化为 int 型数据时，就发生了 NumberFormatException 异常。

以下是一个描述学生基本信息的类 Student.java。

Student.java
```
    package tom.jiafei;
    public class Student {
       String name=null;
       long number;
       double height,weight;
       public String getName() {
          return name;
       }
       public void setName(String newName) {
          name=newName;
       }
       public long getNumber() {
          return number;
       }
       public void setNumber(long newNumber) {
```

```
            number=newNumber;
        }
        public double getHeight() {
            return height;
        }
        public void setHeight(double newHeight) {
            height=newHeight;
        }
        public double getWeight() {
            return weight;
        }
        public void setWeight(double newWeight) {
            weight=newWeight;
        }
    }
```

【例 5-5】 JSP 页面文件 student.jsp 保存在 Web 服务目录 chapter5 中,该页面中使用 Book 类创建两个 bean,其有效范围是 page。在 JSP 页面中使用动作标记设置、获取这些 bean 的属性。效果如图 5.6 所示。

姓名	学号	身高(米)	体重（公斤）
王小林	2007001	1.78	77.87
李四	2007002	1.66	62.65

图 5.6 使用 setProperty 标记设置 bean 属性值

student.jsp
```
<%@ page contentType="text/html;Charset=GB2312" %>
<%@ page import="tom.jiafei.*"%>
<HTML><BODY bgcolor=cyan>
    <jsp:useBean id="wangxiaolin" class="tom.jiafei.Student" scope="page" />
    <jsp:useBean id="li" class="tom.jiafei.Student" scope="page" />
    <jsp:setProperty name="wangxiaolin" property="name" value="王小林" />
    <jsp:setProperty name="li" property="name" value="李四" />
    <jsp:setProperty name="wangxiaolin" property="number" value="2007001" />
    <jsp:setProperty name="li" property="number" value="2007002" />
    <jsp:setProperty name="wangxiaolin" property="height" value="<%=1.78%>" />
    <jsp:setProperty name="li" property="height" value="<%=1.66%>" />
    <jsp:setProperty name="wangxiaolin" property="weight" value="77.87" />
    <jsp:setProperty name="li" property="weight" value="62.65" />
<TABLE border=1>
    <TR> <TH>姓名</TH> <TH>学号</TH>  <TH>身高(米)</TH> <TH>体重（公斤）</TH> </TR>
    <TR> <TD><jsp:getProperty  name= "wangxiaolin" property="name" /></TD>
         <TD><jsp:getProperty  name= "wangxiaolin" property="number" /></TD>
         <TD><jsp:getProperty  name= "wangxiaolin" property="height" /></TD>
```

```
          <TD><jsp:getProperty   name= "wangxiaolin" property="weight" /></TD>
       </TR>
       <TR> <TD><jsp:getProperty   name= "li" property="name" /></TD>
          <TD><jsp:getProperty   name= "li" property="number" /></TD>
          <TD><jsp:getProperty   name= "li" property="height" /></TD>
          <TD><jsp:getProperty   name= "li" property="weight" /></TD>
       </TR>
    </TABLE>
</BODY></HTML>
```

2. 通过 HTTP 表单的参数的值来设置 bean 的相应属性的值

使用 setProperty 设置 bean 属性值的第 2 种方式是：通过 HTTP 表单的参数的值来设置 bean 的相应属性的值，要求表单参数名字必须与 bean 属性的名字相同，Tomcat 服务器会自动将字符串转换为 bean 属性的类型。这种方式的 setProperty 标记的格式如下：

```
<jsp:setProperty   name="bean 的名字"   property="*" />
```

上述格式不用再具体指定 bean 属性的值将对应表单中哪个参数指定的值，系统会自动根据名字进行般配对应。

【例 5-6】 继续使用 Student 创建 bean，用户可以通过表单来指定 bean 的属性值。

studentForm1.jsp

```
<%@ page contentType="text/html;Charset=GB2312" %>
<%@ page import="tom.jiafei.*"%>
<HTML><BODY bgcolor=cyan>
<BODY ><FONT size=1>
<FORM action="" Method="post" >
<P>输入学生的姓名：
<Input type=text name="name">
<P>输入学生的学号：
<Input type=text name="number">
<P>输入学生的身高：
<Input type=text name="height">
<P>输入学生的体重：
<Input type=text name="weight">
<Input type=submit value="提交">
</FORM>
   <jsp:useBean   id="wangxiaolin" class="tom.jiafei.Student" scope="page" />
   <jsp:setProperty   name= "wangxiaolin" property="*" />
<TABLE border=1>
   <TR>
     <TH>姓名</TH> <TH>学号</TH>   <TH>身高(米)</TH> <TH>体重（公斤）</TH>
   </TR>
   <TR>
     <TD><jsp:getProperty   name= "wangxiaolin" property="name" /></TD>
     <TD><jsp:getProperty   name= "wangxiaolin" property="number" /></TD>
```

```
        <TD><jsp:getProperty   name= "wangxiaolin" property="height" /></TD>
        <TD><jsp:getProperty   name= "wangxiaolin" property="weight" /></TD>
    </TR>
 </TABLE>
 </BODY></HTML>
```

注：只有提交了与该 bean 相对应的表单后，相应的 jsp:setProperty 标记才被执行。

3. 通过 request 的参数的值来设置 bean 的相应属性的值

使用 setProperty 设置 bean 属性值的第 3 种方式是：通过 request 的参数的值来设置 bean 的相应属性的值，要求 request 参数名字必须与 bean 属性的名字相同，Tomcat 服务器会自动将 request 获取的字符串数据类型转换为 bean 相应的属性的类型。其格式如下：

    <jsp:setProperty   name= "bean 的名字" property="属性名" param= "参数名" />

【**例 5-7**】 使用 request 参数设置 bean 的属性的值。

studentForm2.jsp
```
        <%@ page contentType="text/html;Charset=GB2312" %>
        <%@ page import="tom.jiafei.*"%>
        <HTML><BODY bgcolor=cyan>
        <BODY ><FONT size=1>
        <FORM action="" Method="post" >
        <P>输入学生的姓名:
        <Input type=text name="name">
        <P>输入学生的学号:
          <Input type=text name="number">
        <P>输入学生的身高:
          <Input type=text name="height">
        <P>输入学生的体重:
          <Input type=text name="weight">
          <Input type=submit value="提交">
          </FORM>
          <jsp:useBean   id="wangxiaolin" class="tom.jiafei.Student" scope="page" />
          <jsp:setProperty   name= "wangxiaolin" property="name" param="name" />
          <jsp:setProperty   name= "wangxiaolin" property="number" param="number" />
        <TABLE border=1>
          <TR>
            <TH>姓名</TH> <TH>学号</TH>   <TH>身高(米)</TH> <TH>体重（公斤）</TH>
          </TR>
          <TR>
            <TD><jsp:getProperty name= "wangxiaolin"   property="name"   /></TD>
            <TD><jsp:getProperty name= "wangxiaolin"   property="number"   /></TD>
            <TD><jsp:getProperty name= "wangxiaolin"   property="height"   /></TD>
            <TD><jsp:getProperty name= "wangxiaolin"   property="weight"   /></TD>
          </TR>
```

    </TABLE>
    </BODY></HTML>

注：只有提交了与该 bean 相对应的表单后，相应的 jsp:setProperty 标记才被执行。

5.3 bean 的辅助类

通过上面的学习，我们已经知道怎样使用一个简单的 bean。有时在写一个 bean 的时候，可能需要自己编写的其他类，那么只要将这些类和创建 bean 的类写在一个 Java 源中即可，但必须按将源文件编译后产生的全部字节码文件复制到相应的目录中（见 5.1.2 节）。

【例 5-8】使用一个 bean —— ListFile，来列出 JSP 页面所在目录中特定扩展名的文件。在写 bean 的类文件 ListFile 时候，需要一个实现 FilenameFilter 接口的辅助类 FileName，该类可以帮助我们的 bean 列出指定扩展名的文件。

ListFile.java
```java
package tom.jiafei;
import java.io.*;
public class ListFile {
    String extendsName,path="";
    StringBuffer allFileName;
    public ListFile() {
        allFileName=new StringBuffer();
    }
    public void setExtendsName(String s) {
        extendsName=s;
    }
    public String getExtendsName() {
        return extendsName;
    }
    public void setPath(String path) {
        this.path=path;
    }
    public String getPath() {
        return path;
    }
    public StringBuffer getAllFileName() {
        if(path.length()>0) {
            File dir=new File(path);
            FileName wantName=new FileName(extendsName);
            String fileName[]=dir.list(wantName);
            for(int i=0;i<fileName.length;i++) {
                allFileName.append("<BR>"+fileName[i]);
            }
        }
    }
```

```
            return allFileName;
        }
    }
    class FileName implements FilenameFilter {
        String str=null;
        FileName (String s) {
            str="."+s;
        }
        public   boolean accept(File dir,String name) {
            return name.endsWith(str);
        }
    }
```

上述 Java 源文件编译通过后,会生成两个字节码文件 ListFile.class 和 FileName.class。需要将这两个字节码文件放在 chapter5\WEB-INF\classes\tom\jiafei 中。

【例 5-9】 客户机通过表单设置 bean 的 extendsName 属性的值,bean 列出所指定目录中具有特定扩展名的文件。(效果如图 5.7 所示)。

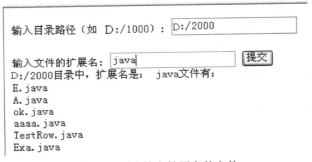

图 5.7 列出特定扩展名的文件

listfile.jsp

```
<%@ page contentType="text/html;charset=GB2312" %>
<%@ page import="tom.jiafei.ListFile" %>
<%@ page import="java.io.File" %>
<HTML><BODY ><FONT size=2>
<FORM action="" Method="post" >
<P>输入目录路径（如 D:/1000）：
<Input type=text name="path" value="D:/2000">
<P>输入文件的扩展名：
<Input type=text name="extendsName">
<Input type=submit value="提交">
  <jsp:useBean   id="file" class="tom.jiafei.ListFile" scope="page" />
  <jsp:setProperty   name= "file" property="*" />
  <BR>
  <jsp:getProperty   name= "file" property="path" />目录中,扩展名是:
   <jsp:getProperty   name= "file" property="extendsName" />文件有:
```

```
<jsp:getProperty   name= "file" property="allFileName" />
</BODY></HTML>
```

5.4 举例

本节通过几个例子来进一步熟悉掌握 bean 的使用。

5.4.1 三角形 bean

编写一个 JSP 页面，该页面提供一个表单，用户可以通过表单将三角形三边的长度提交给该页面。用户提交表单后，JSP 页面将计算三角形面积的任务交给一个 bean 去完成。

创建 bean 的源文件如下：

Triangle.java

```java
package tom.jiafei;
public class Triangle {
    double sideA=-1,sideB=-1,sideC=-1, area=-1;
    boolean triangle;
    public void setSideA(double a) {
        sideA=a;
    }
    public double getSideA() {
        return sideA;
    }
    public void setSideB(double b) {
        sideB=b;
    }
    public double getSideB() {
        return sideB;
    }
    public void setSideC(double c) {
      sideC=c;
    }
    public double getSideC() {
        return sideC;
    }
    public double getArea() {
        double p=(sideA+sideB+sideC)/2.0;
        if(triangle)
            area=Math.sqrt(p*(p-sideA)*(p-sideB)*(p-sideC));
        return area;
    }
    public boolean isTriangle() {
        if(sideA<sideB+sideC&&sideB<sideA+sideC&&sideC<sideA+sideB)
```

```
                triangle=true;
        else
                triangle=false;
        return triangle;
    }
}
```

JSP 页面如下（效果如图 5.8 所示）：

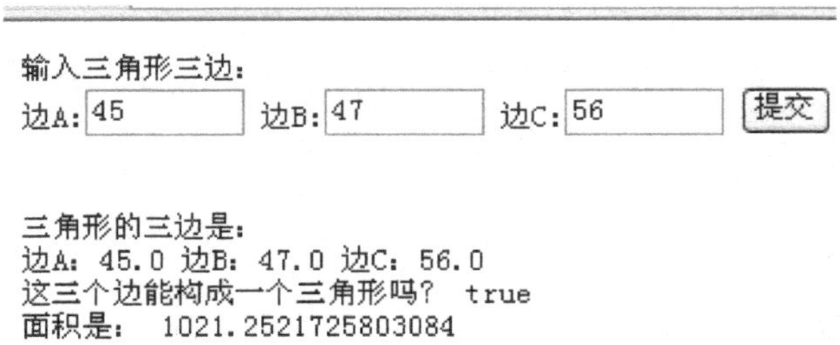

输入三角形三边：
边A: 45   边B: 47   边C: 56   [提交]

三角形的三边是：
边A: 45.0 边B: 47.0 边C: 56.0
这三个边能构成一个三角形吗？ true
面积是： 1021.2521725803084

图 5.8 计算三角形面积

triangle.jsp
```
<%@ page contentType="text/html;Charset=GB2312" %>
<%@ page import="tom.jiafei.Triangle" %>
<jsp:useBean id="triangle" class="tom.jiafei.Triangle" scope="page" />
<HTML><BODY ><FONT size=2>
<FORM action="" Method="post" >
   <P>输入三角形的三条边：
   <BR>边 A:<Input type=text name="sideA" value=0>
       边 B:<Input type=text name="sideB" value=0>
       边 C:<Input type=text name="sideC" value=0>
   <Input type=submit value="提交">
</FORM>
<jsp:setProperty   name= "triangle" property="*" />
<P>三角形的三条边是：
<BR>边 A： <jsp:getProperty   name="triangle" property="sideA" />
    边 B： <jsp:getProperty   name="triangle" property="sideB" />
    边 C： <jsp:getProperty   name="triangle" property="sideC" />
<P>这三个边能构成一个三角形吗？
<jsp:getProperty   name= "triangle" property="triangle" />
<P>面积是：<jsp:getProperty   name= "triangle" property="area" />
</FONT></BODY></HTML>
```

5.4.2 四则运算 bean

编写一个 JSP 页面，该页面提供一个表单，用户可以通过表单输入两个数和四则运算符号提交给该页面。用户提交表单后，该页面将计算任务交给一个 bean 去完成。

创建 bean 的源文件如下：

ComputerBean.java
```
    package tom.jiafei;
```

```java
public class ComputerBean {
    double numberOne,numberTwo,result;
    String operator="+";
    public void setNumberOne(double n) {
        numberOne=n;
    }
    public double getNumberOne() {
        return numberOne;
    }
    public void setNumberTwo(double n) {
        numberTwo=n;
    }
    public double getNumberTwo() {
        return numberTwo;
    }
    public void setOperator(String s) {
        operator=s.trim();;
    }
    public String getOperator() {
        return operator;
    }
    public double getResult() {
        if(operator.equals("+")) {
            result=numberOne+numberTwo;
        }
        else if(operator.equals("-")) {
            result=numberOne-numberTwo;
        }
        else if(operator.equals("*")) {
            result=numberOne*numberTwo;
        }
        else if(operator.equals("/")) {
            result=numberOne/numberTwo;
        }
        return result;
    }
}
```

JSP 页面如下（效果如图 5.9 所示）：

图 5.9 四则运算

computer.jsp
```
<%@ page contentType="text/html;Charset=GB2312" %>
<%@ page import="tom.jiafei.CalendarBean" %>
<HTML><BODY bgcolor=cyan><FONT size=4>
  <jsp:useBean id="computer" class="tom.jiafei.ComputerBean" scope="request"/>
  <FORM action="" method=post name=form>
     输入两个数：
  <Input type=text name="numberOne" value=0 size=8>
  <Input type=text name="numberTwo" value=0 size=8>
     选择运算符号：
     <Select name="operator">
        <Option value="+">+(加)
        <Option value="-">-（减）
        <Option value="*">*（乘）
        <Option value="/">/（除）
     </Select>
  <BR>
  <Input type="submit" value="提交你的选择" name="submit">
  </FORM>
     <jsp:setProperty   name="computer" property="*"/>
     <jsp:getProperty name="computer" property="numberOne"/>
     <jsp:getProperty name="computer" property="operator"/>
     <jsp:getProperty name="computer" property="numberTwo"/>=
     <jsp:getProperty name="computer" property="result"/>
</BODY></HTML>
```

5.4.3 猜数字 bean

当用户访问 getNumber.jsp 页面时，随机给用户一个 1~100 之间的整数，然后用户使用该页面提供的表单输入自己的猜测，并提交给 guess.jsp 页面。guess.jsp 页面使用一个 bean 来处理用户的猜测，判断用户的猜测是否正确。

创建 bean 的源文件如下：

GuessNumber.java
```
package tom.jiafei;
public class GuessNumber {
    int answer=0,                       //待猜测的整数
        guessNumber=0,                  //用户的猜测
        guessCount=0;                   //用户猜测的次数
    String result=null;
    boolean right=false;
    public void setAnswer(int n) {
        answer=n;
        guessCount=0;
    }
```

```java
        public int    getAnswer() {
            return answer;
        }
        public void    setGuessNumber(int n) {
            guessNumber=n;
            guessCount++;
            if(guessNumber==answer) {
                result="恭喜，猜对了";
                right=true;
            }
            else if(guessNumber>answer) {
                result="猜大了";
                right=false;
            }
            else if(guessNumber<answer) {
                result="猜小了";
                right=false;
            }
        }
        public int getGuessNumber() {
            return guessNumber;
        }
        public int getGuessCount() {
            return guessCount;
        }
        public String getResult() {
            return result;
        }
        public boolean isRight() {
            return right;
        }
    }
```

JSP 页面如下：

getNumber.jsp（效果如图 5.10 所示）

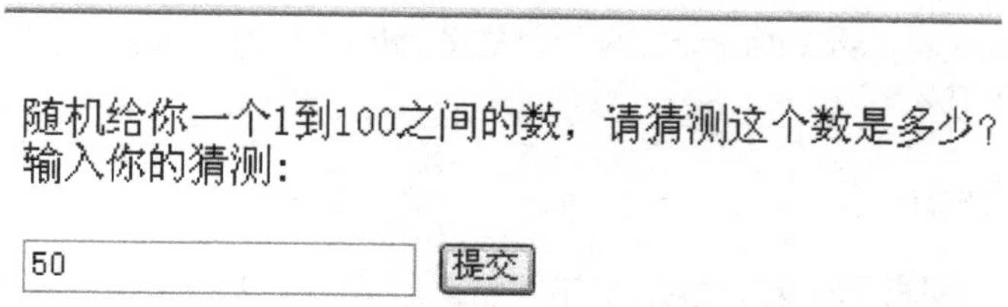

图 5.10　获得一个随机数

```jsp
<%@ page contentType="text/html;charset=GB2312" %>
<%@ page import="tom.jiafei.GuessNumber" %>
<HTML><BODY>
<% int n=(int)(Math.random()*100)+1;%>
```

```
<jsp:useBean  id="guess" class="tom.jiafei.GuessNumber" scope="session" />
<jsp:setProperty  name= "guess" property="answer" value="<%=n%>" />
<p>随机给你一个 1 到 100 之间的数,请猜测这个数是多少?
<% String str=response.encodeRedirectURL("guess.jsp"); %>
<FORM action="<%=str%>" method=post >
<BR>输入你的猜测: <Input type=text name="guessNumber">
<Input type=submit value="提交">
</FORM></BODy>
```

guess.jsp(效果如图 5.11 所示)

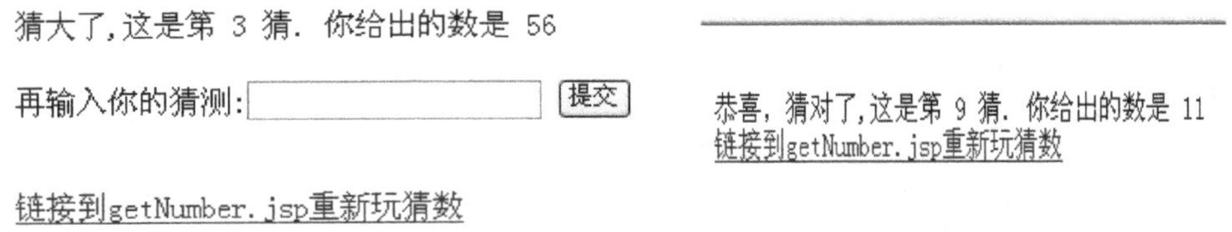

图 5.11 猜数

```
<%@ page contentType="text/html;charset=GB2312" %>
<%@ page import="tom.jiafei.GuessNumber" %>
<% String   strGuess=response.encodeRedirectURL("guess.jsp"),
            strGetNumber=response.encodeRedirectURL("getNumber.jsp");   %>
<HTML><BODY>
<jsp:useBean  id="guess" class="tom.jiafei.GuessNumber" scope="session" />
<jsp:setProperty  name= "guess"  property="guessNumber" param="guessNumber" />
<BR> <jsp:getProperty  name= "guess"  property="result" />,这是第
     <jsp:getProperty name="guess" property="guessCount" /> 猜.
     你给出的数是 <jsp:getProperty  name= "guess"  property="guessNumber" />
     <% if(guess.isRight()==false) {     %>
          <FORM action="<%=strGuess%>" method=post >
               再输入你的猜测: <Input type=text name="guessNumber">
               <Input type=submit value="提交">
          </FORM>
   <%  }  %>
<BR><A href="<%=strGetNumber%>">链接到 getNumber.jsp 重新玩猜数</A>
</BODy></HTML>
```

5.4.4 时间 bean

页面经常需要显示时间有关的数据,如年份、日期等。JSP 页面用一个 bean 负责处理和时间有关的数据,JSP 页面只需调用这样的 bean。

创建 bean 的源文件如下:

ShowCalendar.java
```
    package tom.jiafei;
    import java.util.*;
    public class ShowCalendar {
```

```java
Calendar   calendar = null;
int year,dayOfMonth,dayOfYear,weekOfYear,
weekOfMonth,dayOfWeek,hour,minute,second;
String   day,date,time;
public ShowCalendar(){
    calendar = Calendar.getInstance();
    Date time = new Date();
    calendar.setTime(time);
}
public int getYear(){                             //获取年份
    return calendar.get(Calendar.YEAR);
}
public String getMonth() {                        //获取月，进行格式处理
    int m=1+calendar.get(Calendar.MONTH);
    String months[]={ "1", "2", "3", "4", "5", "6", "7", "8", "9", "10", "11", "12" };
    if (m>12)
        return "0";
    return months[m-1];
}
public String getDay() {                          //获取星期几,进行格式处理
    int n =getDayOfWeek();
    String days[]={"日","一","二","三","四","五","六"};
    if(n>7)
        return "星期？ ";
    return days[n];
}
public String getDate(){                          //获取：年、月、日
    return getYear()+ "/" + getMonth()+"/"+getDayOfMonth();
}
public String getTime() {                         //获取：时：分：秒
    return getHour() + ":" + getMinute() + ":" + getSecond();
}
public int getDayOfMonth() {                      //获取当前时间是一月中的哪一天
    return calendar.get(Calendar.DAY_OF_MONTH);
}
public int getDayOfYear() {                       //获取当前时间是一年中的哪一天
    return calendar.get(Calendar.DAY_OF_YEAR);
}
public int getWeekOfYear() {                      //获取当前时间是一年中的哪个星期
    return calendar.get(Calendar.WEEK_OF_YEAR);
}
public int getWeekOfMonth() {                     //获取当前时间是一年中的哪个星期
    return calendar.get(Calendar.WEEK_OF_MONTH);
}
public int getDayOfWeek() {                       //获取当前时间是一周中的哪一天
    return calendar.get(Calendar.DAY_OF_WEEK)-1;
}
```

```java
        public int getHour() {                              //获取小时
            return calendar.get(Calendar.HOUR_OF_DAY);
        }
        public int getMinute() {                            //获取分钟
            return calendar.get(Calendar.MINUTE);
        }
        public int getSecond() {                            //获取秒
            return calendar.get(Calendar.SECOND);
        }
    }
```

JSP 页面如下：

time.jsp（效果如图 5.12 所示）

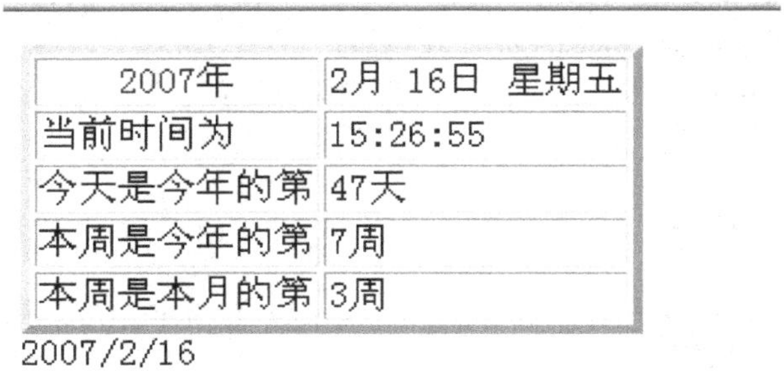

图 5.12 显示时间

```jsp
<%@ page contentType="text/html;Charset=GB2312" %>
<%@ page import="tom.jiafei.ShowCalendar" %>
<HTML><BODY>
<jsp:useBean id="clock" class="tom.jiafei.ShowCalendar" scope="page" />
<TABLE border=4>
<TR>
  <TD align="center">
      <FONT color="blue"><jsp:getProperty name="clock" property="year"/></FONT>年</TD>
  <TD><jsp:getProperty name="clock" property="month"/>月
      <jsp:getProperty name="clock" property="dayOfMonth"/>日
      星期<jsp:getProperty name="clock" property="day"/>
  </TD>
</TR>
<TR>
  <TD>当前时间为</td>
  <TD><jsp:getProperty name="clock" property="time"/></TD>
</TR>
<TR>
  <TD>今天是今年的第</TD>
  <TD><jsp:getProperty name="clock" property="dayOfYear"/>天</TD>
</TR>
<TR>
  <TD>本周是今年的第</TD>
```

```
    <TD><jsp:getProperty name="clock" property="weekOfYear"/>周</TD>
 </TR>
 <TR>
   <TD>本周是本月的第</TD>
   <TD><jsp:getProperty name="clock" property="weekOfMonth"/>周</TD>
 </TR>
</TABLE>
<jsp:getProperty name="clock" property="date"/>
</BODY></HTML>
```

5.4.5 日历 bean

JSP 页面可能需要显示某月的日历，那么 JSP 应当用一个 bean 负责处理时间数据，JSP 页面只需调用这样的 bean。

创建 bean 的源文件如下：

CalendarBean.java

```java
    package tom.jiafei;
    import java.util.*;
    public class CalendarBean {
        String   calendar=null;
        int year=-1,month=-1;
        public void setYear(int year) {
            this.year=year;
        }
        public int getYear() {
           return year;
        }
        public void setMonth(int month) {
            this.month=month;
        }
        public int getMonth() {
            return month;
        }
        public String getCalendar(){
            StringBuffer buffer=new StringBuffer();
            Calendar rili=Calendar.getInstance();
            //将日历翻到 year 年 month 月 1 日，0 表示一月，依次类推，11 表示 12 月
            rili.set(year,month-1,1);
            //获取 1 日是星期几（get 方法返回的值是 1 表示星期日，返回的值是 7 表示星期六）
            int 星期几=rili.get(Calendar.DAY_OF_WEEK)-1;
            int day=0;
            if(month==1||month==3||month==5||month==7||month==8||month==10||month==12) {
                day=31;
            }
            if(month==4||month==6||month==9||month==11) {
```

```
            day=30;
        }
        if(month==2) {
            if(((year%4==0)&&(year%100!=0))||(year%400==0)) {
                day=29;
            }
            else {
                day=28;
            }
        }
        String a[]=new String[42];                //存放号码的一维数组
        for(int i=0;i<星期几;i++) {
            a[i]="**";
        }
        for(int i=星期几,n=1;i<星期几+day;i++) {
            a[i]=String.valueOf(n) ;
            n++;
        }
        for(int i=星期几+day,n=1;i<42;i++) {
            a[i]="**" ;
        }
        //用表格显示数组
        buffer.append("<table border=1>");
        buffer.append("<tr>")   ;
        String weekday[]={"星期日","星期一","星期二","星期三","星期四","星期五","星期六"};
        for(int k=0;k<7;k++) {
            buffer.append("<td>"+weekday[k]+"</td>");
        }
        buffer.append("</tr>") ;
        for(int k=0;k<42;k=k+7) {
            buffer.append("<tr>")   ;
            for(int j=k;j<Math.min(7+k,42);j++) {
                buffer.append("<td>"+a[j]+"</td>");
            }
            buffer.append("</tr>") ;
        }
        buffer.append("</table>");
        calendar=new String(buffer);
        if(year!=-1) {
            return calendar;
        }
        else {
            return "选择一个年份和月份单击"提交"按钮";
        }
    }
```

}
JSP 页面如下:
showcalendar.jsp（效果如图 5.13 所示）

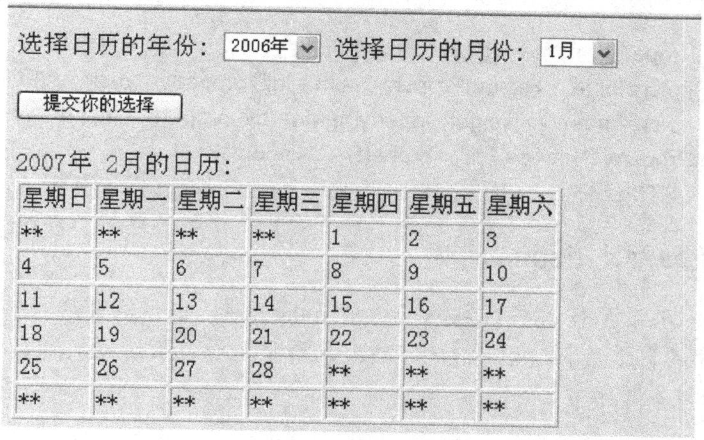

图 5.13　显示日历

```
<%@ page contentType="text/html;Charset=GB2312" %>
<%@ page import="tom.jiafei.CalendarBean" %>
<HTML><BODY bgcolor=cyan><FONT size=4>
   <jsp:useBean id="rili" class="tom.jiafei.CalendarBean" scope="request">
   </jsp:useBean>
<FORM action="" method=post name=form>
   选择日历的年份:
   <Select name="year">
        <Option value="2006">2006 年
        <Option value="2007">2007 年
        <Option value="2008">2008 年
        <Option value="2009">2009 年
        <Option value="2010">2010 年
   </Select>
   选择日历的月份:
   <Select name="month">
        <Option value="1">1 月
        <Option value="2">2 月
        <Option value="3">3 月
        <Option value="4">4 月
        <Option value="5">5 月
        <Option value="6">6 月
        <Option value="7">7 月
        <Option value="8">8 月
        <Option value="9">9 月
        <Option value="10">10 月
        <Option value="11">11 月
         <Option value="12">12 月
```

```
        </Select>
    <BR><BR>
        <Input type="submit" value="提交你的选择" name="submit">
    </FORM>
        <jsp:setProperty    name="rili" property="*"/>
        <FONT color="blue"><jsp:getProperty name="rili" property="year"/></FONT>年
        <FONT color="green"><jsp:getProperty name="rili" property="month"/></FONT>月的日历:
        <jsp:getProperty    name= "rili"   property="calendar" />
    </BODY></HTML>
```

5.4.6 播放幻灯片 bean

幻灯片存放在当前 Web 服务目录的子目录 image 中，每张幻灯片是扩展名为 .jpg 的图像文件（图像名字不能含有汉字和空格）。JSP 页面调用一个负责播放幻灯片的 bean。

创建 bean 的源文件如下：

PlaySlide.java

```java
        package tom.jiafei;
        import java.io.*;
        public class PlaySlide {
            int count=0,max;
            String pictureName[],playNext,playPrevious;
            public PlaySlide() {
                File dir=new File("D:/apache-tomcat-5.5.20/webapps/chaper5/image");
                pictureName=dir.list();
                max=pictureName.length;
            }
            public String getPlayNext() {
                playNext=new String("<image src=image/"+pictureName[count%max]+
                                                " width=120 height=100></image>") ;
                count++;
                if(count>max)
                    count=1;
                return playNext;
            }
            public String getPlayPrevious() {
                playPrevious=new String("<image src=image/"+pictureName[count%max]+
                                                " width=120 height=100></image>");
                count--;
                if(count<0)
                    count=max+1;
                return playPrevious;
            }
        }
```

JSP 页面如下：

play.jsp（效果如图 5.14 所示）

图 5.14　播放幻灯片

```
<%@ page contentType="text/html;Charset=GB2312" %>
<%@ page import="tom.jiafei.PlaySlide" %>
<jsp:useBean id="play" class="tom.jiafei.PlaySlide" scope="session" />
<HTML> <H4>单击"上一张"或"下一张"按钮观看幻灯片
<BODY ><FONT size=3>
  <FORM action="" method=post>
  <BR><Input type=submit name="ok" value="上一张">
      <Input type=submit name="ok" value="下一张">
  </FORM>
  <% String str=request.getParameter("ok");
     if(str==null) {
         str="上一张";
     }
     if(str.equals("上一张")) {
       %>
         <jsp:getProperty   name= "play"   property="playPrevious" />
       <%
     }
     if(str.equals("下一张")) {
  %>     <jsp:getProperty   name="play"   property="playNext" />
  <%
     }
  %>
</FONT></BODY></HTML>
```

习 题 5

1. 假设 Web 服务目录 Dalian 中的 JSP 页面要使用一个 bean，创建该 bean 的字节码文件应当怎样保存？
2. tom.jiafei.Circle 是创建 bean 的类，下列哪个标记是正确创建会话周期 bean 的标记？
   （1）<jsp:useBean id="circle" class="tom.jiafei.Circle" scope="page" />

（2）<jsp:useBean id="circle" class="tom.jiafei.Circle" scope="request" />
（3）<jsp:useBean id="circle" class="tom.jiafei.Circle" scope="session" />
（4）<jsp:useBean id="circle" type="tom.jiafei.Circle" scope="session" />

3. 设创建 bean 的类有一个 int 型的属性 number，下列哪个方法是设置该属性值的正确方法？

（1）
```
public void setNumber(int n) {
    number=n;
}
```

（2）
```
void setNumber(int n) {
    number=n;
}
```

（3）
```
public void SetNumber(int n) {
    number=n;
}
```

（4）
```
public void Setnumber(int n) {
    number=n;
}
```

4. 编写一个 JSP 页面，该页面提供一个表单，用户可以通过表单输入圆的半径提交给本页面，JSP 页面将计算圆的面积和周长之任务交给一个 bean 去完成。

5. 编写 2 个 JSP 页面 a.jsp 和 b.jsp。a.jsp 页面提供一个表单，用户可以通过表单输入矩形的 2 个边长，然后提交给 b.jsp 页面。b.jsp 调用一个 bean 去完成计算矩形面积的任务。

6. 改进 5.4.2 节中的猜数字游戏，使其能显示用户猜数所用时间。

第 6 章

JSP 中的文件操作

本章导读

✪ 知识点：掌握在 JSP 中怎样使用 Java 的输入、输出流实现有关的文件操作。学习使用 JSP+JavaBean 的设计模式，将有关文件的读写指派给 JavaBean。
✪ 重点：学习使用 java.io 包中的类。
✪ 难点：文件的上传与下载。
✪ 关键实践：编写 JSP 页面，实现文件的上传与下载。

在设计一个 Web 应用时，经常会涉及文件操作，如将客户提交的信息以文件的格式保存到服务器端、将服务器上的文件内容显示给客户、提供文件上传与下载功能等。

JSP 通过 Java 的输入/输出流来实现文件的读写操作。本章采用 JSP+JavaBean 的设计模式来学习文件的操作，即将有关文件的读写指派给 bean（如图 6.1 所示）。

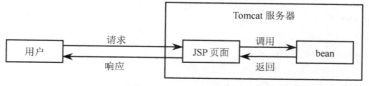

图 6.1 JSP+JavaBean

本章使用的 Web 服务目录是 chapter6，为了使用 bean，在当前 Web 服务目录下建立如下目录结构：chapter6\WEB-INF\classes。然后根据类的包名，在 classes 下建立相应的子目录。为了让 Tomcat 服务器启用上述目录，必须重新启动 Tomcat 服务器。

本章涉及 Java 输入/输出流，为了方便解决中文乱码问题，我们采用的方案是将 page 指令中设置格式中的 charset 的首写字母小写：

<%@ page contentType="text/html;charset=GB2312" %>

6.1 获取文件信息

File 类的对象主要用来获取文件本身的一些信息，如文件所在的目录、文件的长度、文件读写权限等，不涉及对文件的读写操作。

创建一个 File 对象的构造方法有 3 个：

    File(String   filename);
    File(String directoryPath,String filename);
    File(File f, String filename);

其中，filename 是文件名字，directoryPath 是文件的路径，f 是指定成一个目录的文件。

使用 File(String filename)创建文件时，该文件被认为与当前应用程序在同一目录中，由于 Tomcat 服务器在 bin 下启动执行，所以该文件被认为在 Tomcat 安装目录的\bin\中。

经常使用 File 类的下列方法获取文件本身的一些信息：

- public String getName() ——获取文件的名字。
- public boolean canRead() ——判断文件是否是可读的。
- public boolean canWrite() ——判断文件是否可被写入。
- public boolean exists() ——判断文件是否存在。
- public long length() ——获取文件的长度（单位是字节）。
- public String getAbsolutePath() ——获取文件的绝对路径。
- public String getParent() ——获取文件的父目录。
- public boolean isFile() ——判断文件是否是一个正常文件，而不是目录。
- public boolean isDirectroy() ——判断文件是否是一个目录。
- public boolean isHidden() ——判断文件是否是隐藏文件。
- public long lastModified() ——获取文件最后修改的时间（时间是从 1970 年午夜至文件最后修改时刻的毫秒数）。

【例 6-1】 在 JSP 页面调用 bean 获取某些文件的信息。

创建 bean 的源文件：

FilePro.java

```
package tom.jiafei;
import java.io.*;
public class FilePro {
    String filePath,fileName,absolutePath="";
    long length=0,lastModifiedTime=-1;
    boolean exists,canRead;
    public void setFilePath(String s) {
        filePath=s;
    }
    public String getFilePath() {
        return filePath;
    }
    public void setFileName(String s) {
        fileName=s;
```

```
            }
            public String getFileName() {
                return fileName;
            }
            public String getAbsolutePath() {
                if(fileName!=null) {
                    File file=new File(filePath,fileName);
                    absolutePath=file.getAbsolutePath();
                }
                return absolutePath;
            }
            public long getLastModified() {
                if(fileName!=null) {
                    File file=new File(filePath,fileName);
                    lastModifiedTime=file.lastModified();
                }
                return lastModifiedTime;
            }
            public boolean isExits() {
                if(fileName!=null) {
                    File file=new File(fileName);
                    exists=file.exists();
                }
                return exists;
            }
            public boolean isCanRead() {
                if(fileName!=null) {
                    File file=new File(filePath,fileName);
                    canRead=file.canRead();
                }
                return canRead;
            }
            public long getLength() {
                if(fileName!=null) {
                    File file=new File(filePath,fileName);
                    length=file.length();
                }
                return length;
            }
        }
```

JSP 页面：

file.jsp（效果如图 6.2 所示）

```
<%@ page contentType="text/html;charset=GB2312" %>
<%@ page import="tom.jiafei.FilePro" %>
<HTML><BODY bgcolor=cyan><FONT Size=2>
<jsp:useBean id="file" class="tom.jiafei.FilePro" scope="page" />
```

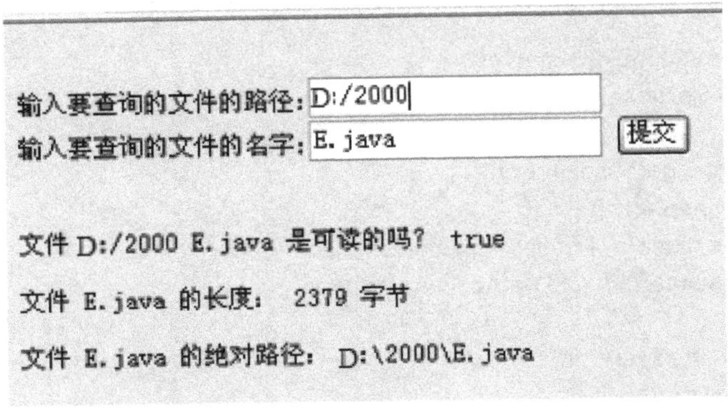

图 6.2 获取文件的有关信息

```
<FORM action="" Method="post" >
   <BR> 输入要查询的文件的路径:<Input type=text name="filePath" value="d:/1000">
   <BR> 输入要查询的文件的名字:<Input type=text name="fileName" value="E.java">
   <Input type=submit value="提交">
</FORM>
  <jsp:setProperty   name= "file"   property="filePath" param="filePath" />
  <jsp:setProperty   name= "file"   property="fileName" param="fileName" />
  <BR> 文件 <jsp:getProperty name="file"   property="filePath"/>
         <jsp:getProperty name="file"   property="fileName"/> 是可读的吗?
         <jsp:getProperty name="file"   property="canRead"/>
  <BR>
  <BR>文件 <jsp:getProperty name="file"   property="fileName"/> 的长度:
         <jsp:getProperty name="file"   property="length"/> 字节
  <BR>
  <BR>文件 <jsp:getProperty name="file"   property="fileName"/>
         的绝对路径:  <jsp:getProperty name="file"   property="absolutePath"/>
</FONT>
</BODY></HTML>
```

6.2 创建、删除 Web 服务目录

在设计一个 Web 应用时，根据具体的应用允许用户在 JSP 页面中调用 bean，在服务器端建立或删除一个目录。

File 对象调用方法如下:

- public boolean mkdir() ——创建一个目录，如果创建成功返回 true，否则返回 false（如果该目录已经存在将返回 false）。
- public boolean delete() ——可以删除当前 File 对象代表的文件或目录，如果 File 对象表示的是一个目录，则该目录必须是一个空目录，删除成功返回 true。

【例 6-2】 在 Tomcat 服务器端创建一个名字是 students 的 Web 服务目录，并列出 Web 服务目录 chapter6 下的全部文件和子目录。

创建 bean 的源文件如下:

FileDir.java

```java
package tom.jiafei;
import java.io.*;
public class FileDir {
    String newWebDirName,oldWebDirName;
    StringBuffer allFileNames;
    public FileDir(){
        allFileNames=new StringBuffer();
    }
    public void setNewWebDirName(String s) {
        newWebDirName=s;
        if(newWebDirName!=null) {
            File dir=new File("D:\\apache-tomcat-5.5.20\\webApps",newWebDirName);
            dir.mkdir();
        }
    }
    public String getNewWebDirName() {
        return newWebDirName;
    }
    public void setOldWebDirName(String s) {
        oldWebDirName=s;
    }
    public String getOldWebDirName() {
        return oldWebDirName;
    }
    public StringBuffer getAllFileNames() {
        if(oldWebDirName!=null) {
            File dir=new File("D:\\apache-tomcat-5.5.20\\webApps",oldWebDirName);
            String a[]=dir.list();
            for(int k=0;k<a.length;k++) {
                allFileNames.append("<BR>"+a[k]);
            }
        }
        return allFileNames;
    }
}
```

JSP 页面如下（效果如图 6.3 所示）：

webdir.jsp

```jsp
<%@ page contentType="text/html;Charset=GB2312" %>
<%@ page import="tom.jiafei.FileDir" %>
<HTML><BODY bgcolor=cyan><FONT Size=2>
<jsp:useBean id="dir" class="tom.jiafei.FileDir" scope="page" />
<FORM action="" Method="post" >
 <BR> 输入新的 Web 服务目录的名字:<Input type=text name="newWebDirName" value="student">
  <BR> 输入已有的 Web 服务目录的名字:
<Input type=text name="oldWebDirName" value="chaper6">
```

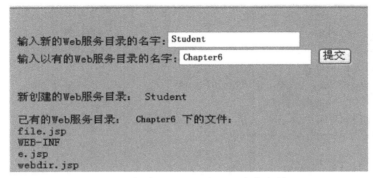

图 6.3 创建目录、列出目录中的文件

```
<Input type=submit value="提交">
</FORM>
  <jsp:setProperty    name= "dir" property="newWebDirName" param="newWebDirName" />
  <jsp:setProperty    name= "dir" property="oldWebDirName" param="oldWebDirName" />
  <BR> 新创建的 Web 服务目录：
<jsp:getProperty name="dir"   property="newWebDirName"/>
  <BR>
  <BR> 已有的 Web 服务目录：
    <jsp:getProperty name="dir"   property="oldWebDirName"/>
    下的文件：
    <jsp:getProperty name="dir"   property="allFileNames"/>
  <BR>
</FONT> </BODY></HTML>
```

6.3 读写文件

Java 的 I/O 流提供一条通道程序，可以把源中的数据送往目的地。我们把输入流的指向称为源，程序从指向源的输入流中读取源中的数据。而输出流的指向是数据要去的一个目的地，程序通过向输出流中写入数据把信息传递到目的地。

6.3.1 读写文件的常用流

1．字节流与字符流

java.io 包提供大量的流类，所有字节输入流类都是 InputStream（输入流）抽象类的子类，而所有字节输出流都是 OutputStream（输出流）抽象类的子类。字节流不能直接操作 Unicode 字符，所以 Java 提供了字符流。由于汉字在文件中占用 2 字节，如果使用字节流，读取不当会出现乱码现象，采用字符流就可以避免这个现象。在 Unicode 字符中，一个汉字被看做一个字符。所有字符输入流类都是 Reader（输入流）抽象类的子类，而所有字符输出流都是 Writer（输出流）抽象类的子类。

输入流 InputStream 类的常用方法如下：

- int read() ——从源中读取单字节的数据，返回字节值（0～255 之间的一个整数），如果未读出字节，就返回 –1。
- int read(byte b[]) ——从源中试图读取 b.length 字节到 b 中,返回实际读取的字节数目。

如果到达文件的末尾，则返回–1。
- int read(byte b[], int off, int len) ——从源中试图读取 len 字节到 b 中，并返回实际读取的字节数目。如果到达文件的末尾，则返回–1，参数 off 指定从字节数组的某个位置开始存放读取的数据。
- void close() ——关闭输入流。
- long skip(long numBytes) ——跳过 numBytes 字节，并返回实际跳过的字节数目。

输入流 Reader 类中的常用方法如下：
- int read() ——从源中读取一个字符，返回一个整数（0～65535 之间的一个整数，Unicode 字符值），如果未读出字符，就返回–1。
- int read(char b[]) ——从源中读取 b.length 个字符到字符数组 b 中，返回实际读取的字符数目。如果到达文件的末尾，则返回–1。
- int read(char b[], int off, int len) ——从源中读取 len 个字符并存放到字符数组 b 中，返回实际读取的字符数目。如果到达文件的末尾，则返回–1。其中，off 参数指定 read 方法从符数组 b 中的什么地方存放数据。
- void close() ——关闭输入流。
- long skip(long numBytes) ——跳过 numBytes 个字符，并返回实际跳过的字符数目。

输出流 OutputStream 类的常用方法如下：
- void write(int n) ——向输出流中写入单字节。
- void write(byte b[]) ——向输出流中写入单字节数组。
- void write(byte b[], int off, int len) ——从给定字节数组中起始于偏移量 off 处取 len 字节写到输出流。
- void close() ——关闭输出流。

输出流 Writer 类中常用方法：
- void write(int n) ——向输出流中写入单字符。
- void write(byte b[])——向输出流中写入单字符数组。
- void write(byte b[], int off, int length) ——从给定字符数组中起始于偏移量 off 处取 len 个字符写到输出流。
- void close()——关闭输出流。

字节流和字符流的 read 方法都是顺序地读取源中的数据，只要不关闭流，每次调用 read 方法就顺序地读取源中其余的内容，直到源的末尾或流被关闭。

字节流和字符流的 writer 方法都是顺序地写数据到目的地，只要不关闭流，每次调用 writer 方法就顺序地向目的地写入内容，直到流被关闭。

2．FileInputStream 与 FileOutputStream 流

FileInputStream 类是从 InputStream 中派生出来的简单输入流类，其所有方法都是从 InputStream 类继承来的。为了创建 FileInputStream 类的对象，用户可以调用它的构造方法：

    FileInputStream(String name)
    FileInputStream(File file)

第一个构造方法使用给定的文件名 name 创建一个 FileInputStream 对象，第二个构造方法使用 File 对象创建 FileInputStream 对象。参数 name 和 file 指定的文件称做输入流的源，

输入流通过调用 read 方法读出源中的数据。

与 FileInputStream 类相对应的类是 FileOutputStream 类。FileOutputStream 提供了基本的文件写入能力。除了从 FileOutputStream 类继承来的方法以外，FileOutputStream 类还有两个常用的构造方法：

  FileOutputStream(String name)
  FileOutputStream(File file)

第一个构造方法使用给定的文件名 name 创建一个 FileOutputStream 对象。第二个构造方法使用 File 对象创建 FileOutputStream 对象。参数 name 和 file 指定的文件称做输出流的目的地，通过向输出流中写入数据把信息传递到目的地。

3．FileReader 与 FileWriter 类

FileReader 类是从 Reader 中派生出来的简单输入类，其所有方法都是从 Reader 类继承来的。为了创建 FileReader 类的对象，用户可以调用它的构造方法。例如：

  FileReader(String name)
  FileReader (File file)

第一个构造方法使用给定的文件名 name 创建一个 FileReader 对象，第二个构造方法使用 File 对象创建 FileReader 对象。参数 name 和 file 指定的文件称为输入流的源，输入流通过调用 read 方法读出源中的数据。

与 FileReader 类相对应的类是 FileWriter 类。FileWriter 类提供了基本的文件写入能力。除了从 FileWriter 类继承来的方法以外，FileWriter 类还有两个常用的构造方法：

  FileWriter(String name)
  FileWriter (File file);

第一个构造方法使用给定的文件名 name 创建一个 FileWriter 对象。第二个构造方法使用 File 对象创建 FileWriter 对象。参数 name 和 file 指定的文件称为输出流的目的地，通过向输出流中写入数据把信息传递到目的地。

创建输入流、输出流对象能发生 IOException 异常，必须在 try、catch 块语句中创建输入、输出流对象。

为了提高读写的效率，FileReader 流经常与 BufferedReader 流配合使用；FileWriter 流经常与 BufferedWriter 流配合使用。BufferedReader 流还可以使用 String readLine()方法读取一行；BufferedWriter 流还可以使用 void write(String s,int off,int length)方法将字符串 s 的一部分写入文件，使用 newLine()方法向文件写入一个行分隔符。

6.3.2 读取文件

【例 6-3】 创建 bean 类，使用输入/输出流，可以列出用户指定的目录中的文件，并读取目录中文件的内容，bean 的 scope 取值为 session。其中有两个 JSP 页面：selectdir.jsp 和 listfile.jsp，selectdir.jsp 页面调用该 bean 选择目录，listfile.jsp 页面调用该 bean 读取相应目录下文件。

创建 bean 的源文件如下：

ReadFile.java

```
package tom.jiafei;
import java.io.*;
```

```java
public class ReadFile {
    String fileDir="c:/",fileName="";
    String listFile,readContent;
    public void setFileDir(String s) {
        fileDir=s;
    }
    public String getFileDir() {
        return fileDir;
    }
    public void setFileName(String s) {                    //设置文件名字属性的值
        fileName=s;
    }
    public String getFileName() {
        return fileName;
    }
    public String getListFile(){                            //列出指定目录中的文件
        File dir=new File(fileDir);
        File file_name[]=dir.listFiles();
        StringBuffer list=new   StringBuffer();
        for(int i=0;i<file_name.length;i++) {
            if ((file_name[i]!=null)&&(file_name[i].isFile())) {
                String temp=file_name[i].toString();
                int n=temp.lastIndexOf("\\");
                temp=temp.substring(n+1);
                list.append(" "+temp);
            }
        }
        listFile=new String(list);
        return listFile;
    }
    public String getReadContent() {                        //读取文件
        try{
            File file=new File(fileDir,fileName);
            FileReader in=new FileReader(file) ;
            BufferedReader inTwo=new BufferedReader(in);
            StringBuffer stringbuffer=new StringBuffer();
            String s=null;
            while ((s=inTwo.readLine())!=null) {
                stringbuffer.append("\n"+s);
            }
            String temp=new String(stringbuffer);
            readContent="<TextArea rows=10 cols=62>"+temp+"</TextArea>";
        }
        catch(IOException e) {
            readContent="<TextArea rows=8 cols=62>"+"请选择一个文件"+"</TextArea>";
        }
```

```
            return readContent;
        }
    }
```
JSP 页面文件如下：

selectdir.jsp（效果如图 6.4 所示）：

图 6.4　选择目录

```
<%@ page contentType="text/html;charset=GB2312" %>
<HTML><BODY ><FONT size=3>
<P>请选择一个目录：
<FORM action="listfile.jsp" method=post>
  <Select name="fileDir" >
    <Option value="D:/1000"> D:/1000
    <Option value="D:/apache-tomcat-5.5.20/webapps/chaper6">Web 服务目录 chaper6
    <Option value="D:/apache-tomcat-5.5.20/webapps/chaper3">Web 服务目录 chaper3
  </Select>
  <Input type=submit value="提交">
</FORM>
</FONT></BODY></HTML>
```

listfile.jsp（效果如图 6.5 所示）

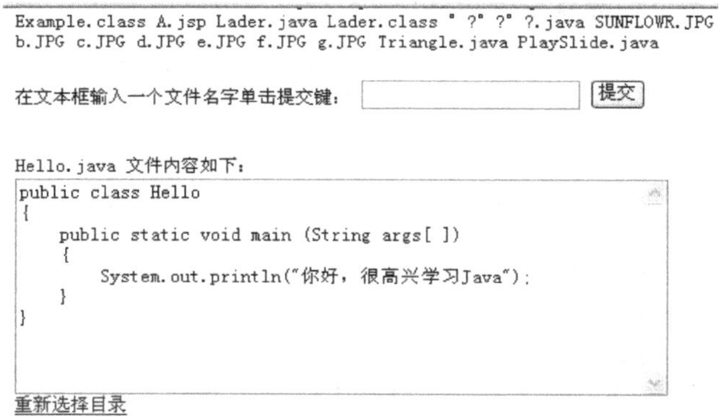

图 6.5　读取文件内容

```
<%@ page contentType="text/html;charset=GB2312" %>
<%@ page import="tom.jiafei.ReadFile" %>
<HTML><BODY ><FONT size=2>
<jsp:useBean id="file" class="tom.jiafei.ReadFile" scope="session" />
<jsp:setProperty  name="file"  property="fileDir" param="fileDir" />
<P>该目录
  <jsp:getProperty  name= "file"  property="fileDir"  />
   有如下文件：
<BR>
  <jsp:getProperty  name= "file"  property="listFile"  />
```

```
<FORM action="" method=post name=form1>
    在文本框输入一个文件名字单击提交键:
    <Input type=text name="fileName">
    <Input type=submit value="提交">
</FORM>
    <jsp:setProperty  name= "file"  property="fileName" param="fileName" />
<BR>
    <jsp:getProperty  name= "file"  property="fileName"/>
    文件内容如下:
<BR>
    <jsp:getProperty  name= "file"  property="readContent"  />
<BR>
    <A href="selectdir.jsp">重新选择目录</A>
</Body></HTML></HTML>
```

6.3.3 按行读取

当读取一个较大的文件时,用户可能希望分行读取该文件。由于 HTTP 协议是一种无状态协议,客户向服务器发出请求(request)然后服务器返回响应(response),连接就被关闭了。也就是说,如果用户每次请求读取 10 行,那么第一次请求会读取文件的前 10 行,当第 2 次请求时读取的仍然是文件的前 10 行。

Tomcat 服务器可以借助 scope 取值为 session 的 bean 来实现分行读取文件,该 bean 建立一个指向该文件的输入流,只要文件没有读取完毕,JSP 页面每次都使用该 bean 的输入流继续读取文件,直到文件被读取完毕。

【例 6-4】 实现分行读取文件,用户可以通过 JSP 页面选择读取 D:/2000 下的某个文件,并可以设置每次读取的行数。

创建 bean 的源文件如下:

ReadByRow.java
```java
package tom.jiafei;
import java.io.*;
public class ReadByRow {
    FileReader inOne;
    BufferedReader inTwo;
    String fileName;
    StringBuffer readMessage;
    int rows=0;
    public ReadByRow() {
        readMessage=new StringBuffer();
    }
    public void setFileName(String s) {
        fileName=s;
        readMessage=new StringBuffer();
        rows=0;
        File f=new File("D:/2000",fileName);
        try{   inOne=new FileReader(f);
               inTwo=new BufferedReader(inOne);
```

```java
            }
            catch(IOException exp) { }
        }
        public String getFileName() {
            return fileName;
        }
        public void setRows(int n) {
            rows=n;
        }
        public int getRows() {
            return rows;
        }
        public StringBuffer getReadMessage() {
            int i=1;
            try{   while(i<=rows) {
                    String str=inTwo.readLine();
                    if(str==null) {
                        inOne.close();
                        inTwo.close();
                        readMessage.append("<BR><font size=4 color=red>文件读取完毕</font>");
                        break;
                    }
                    else {
                        readMessage.append("<BR>"+str);
                    }
                    i++;
                }
            }
            catch(Exception exp) { }
            return readMessage;
        }
    }
}
```

JSP 页面文件如下：

selectFile.jsp

```jsp
<%@ page contentType="text/html;charset=GB2312" %>
<%@ page import="tom.jiafei.ReadByRow" %>
<jsp:useBean id="reader" class="tom.jiafei.ReadByRow" scope="session" />
<HTML><BODY ><FONT size=3>
<BR>
 <FORM action="readByRow.jsp" method=post>
  <BR> 请选择一个文件：
     <Select name="fileName" >
        <Option value="A.java"> A.java
        <Option value="E.java">E.java
        <Option value="B.jsp">B.jsp
     </Select>
  <BR>输入每次读取的行数:
     <Input type=text name="rows" value="0" size=8>
```

```
    <BR><Input type=submit value="提交">
</FORM>
</FONT></BODY></HTML>
```

readByRow.jsp
```
<%@ page contentType="text/html;charset=GB2312" %>
<%@ page import="tom.jiafei.ReadByRow" %>
<jsp:useBean id="reader" class="tom.jiafei.ReadByRow" scope="session" />
<HTML>
<BODY ><FONT size=3>
   <jsp:setProperty  name= "reader"   property="fileName" param="fileName" />
   <jsp:setProperty  name= "reader"   property="rows" param="rows" />
   <BR>正在读取的文件的名字: <jsp:getProperty  name= "reader"   property="fileName" />
   <A href="selectFile.jsp">重新选择文件</A>
   <FORM action="" method=post>
      <Input type=submit value="读取">
   </FORM>
   <jsp:getProperty  name= "reader"   property="readMessage" />
</FONT></BODY></HTML>
```

6.3.4 写文件

【**例 6-5**】 用户可以通过 JSP 页面提供的文本区输入文本，单击【提交】按钮，JSP 页面调用 bean 完成文件的写入操作，该 bean 可以将客户提交的文本内容写入到一个指定目录的文件中。

创建 bean 的源文件如下：

WriteFile.java
```
package tom.jiafei;
import java.io.*;
public class WriterFile {
    String filePath="",
           fileName="",
           fileContent="";
    public void setFilePath(String s) {
        filePath=s;
    }
    public String getFilePath() {
        return filePath;
    }
    public void setFileName(String s) {
        fileName=s;
    }
    public String getFileName() {
        return fileName;
    }
    public void setFileContent(String s) {
      fileContent=s;
```

```
            if(fileName.length()>0&&filePath.length()>0)
               try{   File file=new File(filePath,fileName);
                      FileOutputStream in=new FileOutputStream(file) ;
                      byte bb[]=fileContent.getBytes("ISO-8859-1");
                      in.write(bb);
                      in.close();
                  }
               catch(Exception e){ }
          }
          public String getFileContent() {
             return fileContent;
          }
     }
```

JSP 页面文件如下：

write.jsp
```
<%@ page contentType="text/html;Charset=GB2312" %>
<%@ page import="tom.jiafei.WriterFile" %>
<jsp:useBean id="writeFile" class="tom.jiafei.WriterFile" scope="page" />
<HTML><BODY ><FONT size=3>
<P>请选择一个目录：
<FORM action="" method=post>
  <Select name="filePath" >
     <Option value="D:/2000"> D:/2000
     <Option value="d:/1000">D:/1000
   </Select>
<BR>输入保存文件的名字： <Input type=text name="fileName" >
<BR>输入文件的内容：
<BR>
  <TextArea   name= "fileContent" Rows= "10" Cols= "40">   </TextArea>
<BR><Input type=submit value="提交">
</FORM>
   <jsp:setProperty   name= "writeFile"   property="filePath" param="filePath" />
   <jsp:setProperty   name= "writeFile"   property="fileName" param="fileName" />
   <jsp:setProperty   name= "writeFile"   property="fileContent" param="fileContent" />
<BR>你写文件到目录：
   <jsp:getProperty   name= "writeFile"   property="filePath"   />
<BR>文件的名字是：
   <jsp:getProperty   name= "writeFile"   property="fileName"   />
<BR>文件的内容是：
   <jsp:getProperty   name= "writeFile"   property="fileContent"   />
</FONT></BODY></HTML>
```

6.4 标准化考试

使用网络进行标准化考试是一种常见的考试形式，大部分标准化考试都使用数据库来处理有关数据。数据库可以方便地管理有关数据，却降低了系统的效率。基于文件来管理有关数据，可以提高系统的效率，但要求合理地组织有关数据，以便系统方便地管理数据。

本节将使用 bean 来管理试题文件，JSP 页面调用 bean 实现网络标准化考试。

为了使 bean 方便地处理数据，要求试题文件存放在 D:\2000 中，并且试题文件的第一行必须是全部试题的答案（用来判定考试者的分数）。例如：

```
CDA
1. Do you know ____ he will ride here at 8 tomorrow morning.
   A. when          B. where        C.whether       D.how
2. Could you tell me___ ? I am his old friend.
   A. where does Jim live        B. when will Jim com back
   C. how is Jim                 D. where Jim has gone.
3. Who is ___ girl in your class?
   A. the shortest    B. shorter    C. shortest    D.short
```

试题可以是任何一套标准的测试题，如英语的标准化考试，包括单词测试，阅读理解等，试题中每道题目提供 A、B、C、D 四个选择。客户可以在几套试题中任选一套试题。

【例 6-6】 bean 负责读入试题、判定分数。JSP 页面调用 bean 显示试题，当用户在 JSP 页面提供的文本框中输入全部答案，单击【提交】按钮后，JSP 页面再次调用 bean 来判定用户的得分。

创建 bean 的源文件如下：

Test.java

```java
package tom.jiafei;
import java.io.*;
public class Test {
    String fileName="",                    //存放考题文件名字的字符串
           correctAnswer="",               //存放正确答案的字符串
           testContent="",                 //存放试题的字符串
           selection="" ;                  //客户提交的答案的字符串
    int score=0;                           //考试者的得分
    public void setFileName(String name) {
        fileName=name;
        selection="" ;
        score=0;
    }
    public String getFileName(){
        return fileName;
    }
    public String getCorrectAnswer() {                //读取试题文件的第一行：标准答案
        try {   File f=new File("D:/2000",fileName);
                FileReader in=new FileReader(f);
                BufferedReader buffer=new BufferedReader(in);
                correctAnswer=(buffer.readLine()).trim();   //读取一行,去掉前后空格
                buffer.close();
                in.close();
        }
        catch(Exception e){ }
        if(selection.length()>0)
            return correctAnswer;
```

```
                else
                    return "提交答案后，可以看到正确答案";
        }
        public String getTestContent() {                        //获取试题的内容
            StringBuffer temp=new StringBuffer();
            try {   if(fileName.length()>0) {
                    File f=new File("D:/2000",fileName);
                    FileReader in=new FileReader(f);
                    BufferedReader buffer=new BufferedReader(in);
                    String str=buffer.readLine();               //该行不显示给用户
                    while((str=buffer.readLine())!=null) {      //读出全部题目
                        temp.append("\n"+str);
                    }
                    buffer.close();
                    in.close();
                }
            }
            catch(Exception e) { }
            return "<TextArea rows=15 cols=80>"+new String(temp)+"</TextArea>";
        }
        public void setSelection(String s) {
            selection=s.trim();
        }
        public String getSelection() {
            return selection;
        }
        public int getScore() {
            score=0;
            int length1=selection.length();
            int length2=correctAnswer.length();
            int min=(int)(Math.min(length1,length2));
            int i=0;
            while(i<min) {
                if(selection.charAt(i)==correctAnswer.charAt(i))
                    score++;
                i++;
            }
            return score;
        }
    }
```

JSP 页面文件如下：

test.jsp
```
<%@ page contentType="text/html;charset=GB2312" %>
<%@ page import="tom.jiafei.Test" %>
<HTML><BODY ><FONT size=2>
<jsp:useBean id="test" class="tom.jiafei.Test" scope="session" />
<BR>请选择试题:
```

```
<Form action="" method=post name=form1>
  <Select name="fileName"   value="A.txt">
    <Option value="A.txt"> A.txt
    <Option value="B.txt"> B.txt
    <Option value="C.txt"> C.txt
  </Select>
  <Input type="submit" name="sub" value="确定">
</FORM>
  <jsp:setProperty   name= "test"   property="fileName" param="fileName" />
<BR>试题内容如下：
<BR> <jsp:getProperty   name= "test"   property="testContent"   /> <BR>
<FORM action="" method=post name=form2>
  在文本框输入全部题目的答案，答案之间不允许有空格：
  <BR><Input type=text name="selection" size=80>
  <BR><Input type=submit value="提交">
</FORM>
  <jsp:setProperty   name= "test"   property="selection"   />
<BR>试题的正确答案：   <jsp:getProperty   name= "test"   property="correctAnswer"   />
<BR>您提交的答案：   <jsp:getProperty   name= "test"   property="selection"   />
<BR>您的分数：   <jsp:getProperty   name= "test"   property="score"   />
</BODY></HTML>
```

6.5 文件上传

Web 应用经常提供文件上传功能，本节学习怎样使用 RandomAccessFile 类提供的功能来实现文件上传。

RandomAccessFile 类创建的流与前面的输入流、输出流不同，RandomAccessFile 类既不是输入流类 InputStream 的子类，也不是输出流类 OutputStram 的子类。习惯上，仍然称 RandomAccessFile 类创建的对象为一个流，RandomAccessFile 流的指向既可以作为源，也可以作为目的地。换句话说，当我们想对一个文件进行读写操作时，可以创建一个指向该文件的 RandomAccessFile 流，这样既可以从这个流中读取这个文件的数据，也通过这个流写入数据给这个文件。

RandomAccessFile 类的两个构造方法如下：

- RandomAccessFile(String name, String mode) ——参数 name 用来确定一个文件名，给出创建的流的源（也是流目的地），参数 mode 取"r"（只读）或"rw"（可读写），决定创建的流对文件的访问权。
- RandomAccessFile(File file, String mode) ——参数 file 是一个 File 对象，给出创建的流的源（也是流目的地），参数 mode 取"r"（只读）或"rw"（可读写），决定创建的流对文件的访问权。创建对象时应捕获 IOException 异常。

RandomAccessFile 类中有一个方法 seek(long a)，用来移动 RandomAccessFile 流指向的文件的指针，参数 a 确定文件指针距离文件开头的字节位置。流还可以调用 getFilePointer() 方法获取当前文件的指针的位置（RandomAccessFile 类的一些方法见表 6.1），RandomAccessFile 流对文件的读写比顺序读写的文件输入/输出流更灵活。

表 6.1　RandomAccessFile 类的常用方法

方　　法	描　　述	方　　法	描　　述
close()	关闭文件	getFilePointer()	获取文件指针的位置
length()	获取文件的长度	read()	从文件中读取一个字节的数据
readBoolean()	从文件中读取一个布尔值，0 代表 false，其他值代表 true	readDouble()	从文件中读取一个双精度浮点值（8 字节）
readByte()	从文件中读取一个字节	readChar()	从文件中读取一个字符（2 字节）
readFloat()	从文件中读取一个单精度浮点值（4 字节）	readFully(byte b[])	读 b.length 字节放入数组 b，完全填满该数组
readInt()	从文件中读取一个 int 值（4 字节）	readLine()	从文件中读取一个文本行
readlong()	从文件中读取一个长型值（8 字节）	readShort()	从文件中读取一个短型值（2 字节）
readUTF()	从文件中读取一个 UTF 字符串	seek()	定位文件指针在文件中的位置
setLength(long newlength)	设置文件的长度	skipBytes(int n)	在文件中跳过给定数量的字节
write(byte b[])	写 b.length 个字节到文件	writeBoolean(boolean v)	把一个布尔值作为单字节值写入文件
writeByte(int v)	向文件写入一个字节	writeBytes(String s)	向文件写入一个字符串
writeChar(char c)	向文件写入一个字符	writeChars(String s)	向文件写入一个作为字符数据的字符串
writeDouble(double v)	向文件写入一个双精度浮点值	writeFloat(float v)	向文件写入一个单精度浮点值
writeInt(int v)	向文件写入一个 int 值	writeLong(long v)	向文件写入一个长型 int 值
writeShort(int v)	向文件写入一个短型 int 值	writeUTF(String s)	写入一个 UTF 字符串

JSP 页面提供 File 类型的表单，File 类型的表单可以让用户选择要上传的文件。File 类型表单的格式如下：

```
<FORM action= "接受上传文件的页面"  method= "post"  enctype=" multipart/form-data"
<Input type= "File"  name= "参数名字"  >
</FORM>
```

bean 负责将用户选择的文件上传到服务器。bean 可以让内置对象 request 调用方法 getInputStream() 获得一个输入流，通过这个输入流读入客户上传的全部信息，包括文件的内容和表单域的信息。bean 可以从上传的全部信息中分离出文件的内容，并保存在服务器上。按照 HTTP，文件表单提交的信息中，前 4 行和后 5 行是表单本身的信息，中间部分才是客户提交的文件的内容。bean 通过使用 RandomAccessFile 流获取文件的内容，即去掉表单的信息。首先，bean 将客户提交的全部信息保存为一个临时文件，该文件的名字是客户的 session 对象的 id（不同客户的这个 id 是不同的），然后读取该临时文件的第 2 行，这一行中含有客户上传的文件的名字，获取这个名字，再获取第 4 行结束的位置，以及倒数第 6 行的结束位置，因为这两个位置之间的内容是上传文件的内容，然后将这部分内容存入文件，该文件的名字与客户上传的文件的名字保持一致。最后删除临时文件。

【例 6-7】

创建 bean 的源文件如下：

UpFile.java

```
package tom.jiafei;
import java.io.*;
import javax.servlet.http.*;
```

```java
public class UpFile {
    HttpServletRequest request;
    HttpSession session;
    String upFileMessage="";
    public void setRequest(HttpServletRequest request) {
        this.request=request;
    }
    public void setSession(HttpSession session) {
        this.session=session;
    }
    public String getUpFileMessage() {
        String fileName=null;
        try{  String tempFileName=(String)session.getId();        //客户的 session 的 id
            File f1=new File("D:/apache-tomcat-5.5.20/webapps/chaper6",tempFileName);
            FileOutputStream o=new FileOutputStream(f1);
            InputStream in=request.getInputStream();
            byte b[]=new byte[10000];
            int n;
            while( (n=in.read(b))!=-1) {                           //将客户上传的全部信息存入 f1
                o.write(b,0,n);
            }
            o.close();
            in.close();
            RandomAccessFile random=new RandomAccessFile(f1,"r");
            int second=1;                             //读出 f1 的第 2 行, 析取出上传文件的名字
            String secondLine=null;
            while(second<=2) {
                secondLine=random.readLine();
                second++;
            }
            //获取第 2 行中目录符号'\'最后出现的位置
            int position=secondLine.lastIndexOf('\\');
            //客户上传的文件的名字是:
            fileName=secondLine.substring(position+1,secondLine.length()-1);
            byte   cc[]=fileName.getBytes("ISO-8859-1");
            fileName=new String(cc);
            session.setAttribute("Name",fileName);                 //供 show.jsp 页面使用
            random.seek(0);                                        //再定位到文件 f1 的开头
            //获取第 4 行回车符号的位置
            long   forthEndPosition=0;
            int forth=1;
            while((n=random.readByte())!=-1&&(forth<=4)) {
                if(n=='\n') {
                    forthEndPosition=random.getFilePointer();
                    forth++;
                }
            }
            //根据客户上传文件的名字, 将该文件存入磁盘
            File f2= new File("D:/apache-tomcat-5.5.20/webapps/chaper6",fileName);
            RandomAccessFile random2=new RandomAccessFile(f2,"rw");
            //确定出文件 f1 中包含客户上传的文件的内容的最后位置, 即倒数第 6 行
```

```
                    random.seek(random.length());
                    long endPosition=random.getFilePointer();
                    long mark=endPosition;
                    int j=1;
                    while((mark>=0)&&(j<=6)) {
                        mark--;
                        random.seek(mark);
                        n=random.readByte();
                        if(n=='\n') {
                            endPosition=random.getFilePointer();
                            j++;
                        }
                    }
                    //将 random 流指向文件 f1 的第 4 行结束的位置
                    random.seek(forthEndPosition);
                    long startPoint=random.getFilePointer();
                    //从 f1 读出客户上传的文件存入 f2(读取从第 4 行结束位置和倒数第 6 行之间的内容)
                    while(startPoint<endPosition-1) {
                        n=random.readByte();
                        random2.write(n);
                        startPoint=random.getFilePointer();
                    }
                    random2.close();
                    random.close();
                    f1.delete();                                    //删除临时文件
                    upFileMessage=fileName+" Successfully UpLoad";
                    return upFileMessage;
                }
                catch(Exception exp) {
                    if(fileName!=null) {
                        upFileMessage=fileName+" Fail to UpLoad";
                        return upFileMessage;
                    }
                    else {
                        upFileMessage="";
                        return upFileMessage;
                    }
                }
            }
        }
```

JSP 页面文件如下：

upfile.jsp

```
<%@ page contentType="text/html;charset=GB2312" %>
<%@ page import="tom.jiafei.UpFile" %>
<jsp:useBean id="upFile" class="tom.jiafei.UpFile" scope="session" />
<HTML><BODY> <P>选择要上传的文件： <BR>
  <FORM action="" method="post" enctype="multipart/form-data">
     <Input type=FILE name="boy" size="45">
     <BR> <Input type="submit" name ="g" value="提交">
  </FORM>
```

```
        <%  upFile.setRequest(request);
            upFile.setSession(session);
        %>
        <jsp:getProperty   name="upFile" property="upFileMessage"/>
        <P>如果上传的是图像文件，可单击超链接查看图像：
        <BR><A href="show.jsp"> 查看图像</A>
    </BODY></HTML>
```

show.jsp

```
    <%@ page contentType="text/html;Charset=GB2312" %>
    <jsp:useBean id="upFile" class="tom.jiafei.UpFile" scope="session" />
    <HTML><BODY>
      <% String pic=(String)session.getAttribute("Name");
         out.print(pic);
         out.print("<image src="+pic+">");
      %>
    </BODY></HTML>
```

6.6 文件下载

Tomcat 5.5 服务器提供了方便的下载功能。只需让内置对象 response 调用方法 setHeader，添加下载的头给客户的浏览器即可，浏览器收到该头后就会打开相应的下载对话框。response 调用 setHeade()方法添加下载头的格式如下：

response.setHeader("Content-disposition","attachment;filename="下载的文件名字");

【例 6-8】客户在 JSP 页面选择一个文件，JSP 页面调用 bean 下载所选择的文件。

创建 bean 的源文件如下：

DownLoadFile.java

```
    package tom.jiafei;
    import java.io.*;
    import javax.servlet.http.*;
    public class DownLoadFile {
        HttpServletResponse response;
        String fileName;
        public void setResponse(HttpServletResponse response) {
            this.response=response;
        }
        public String getFileName() {
            return fileName;
        }
        public void setFileName(String s) {
            fileName=s;
            File fileLoad=new File("F:/2000",fileName);
            //客户使用下载文件的对话框:
            response.setHeader("Content-disposition","attachment;filename="+fileName);
            //下载的文件:
            try{
                //读取文件,并发送给用户下载:
                FileInputStream in=new FileInputStream(fileLoad);
                OutputStream o=response.getOutputStream();
```

```
                    int n=0;
                    byte b[]=new byte[500];
                    while((n=in.read(b))!=-1)
                        o.write(b,0,n);
                    o.close();
                    in.close();
                }
                catch(Exception exp){ }
            }
        }
```

JSP 页面文件如下：

downfile.jsp
```
<%@ page contentType="text/html;charset=GB2312" %>
<%@ page import="tom.jiafei.DownLoadFile" %>
<%@ page import="java.io.*" %>
<jsp:useBean id="downFile" class="tom.jiafei.DownLoadFile" scope="page" />
<HTML><BODY> <P>选择要下载的文件：
   <FORM action="">
      <Select name="fileName" >
         <Option value="book.zip">book.zip
         <Option value="A.java">A.java
         <Option value="B.jsp">B.jsp
      </Select>
      <Input type="submit" value="提交你的选择" name="submit">
   </FORM>
   <%  downFile.setResponse(response);
   %>
<jsp:setProperty  name="downFile" property="fileName" param="fileName"/>
</BODY></HTML>
```

习 题 6

1. File 类对象的主要作用是什么？
2. File 类对象怎样获取文件的长度？
3. RandomAccessFile 类创建的流在读写文件时有什么特点？
4. 按行读取文件的关键技术是什么？
5. 编写 2 个 JSP 页面 a.jsp 和 b.jsp。a.jsp 通过表单提交一个 Web 服务目录的名字给 b.jsp，b.jsp 根据 a.jsp 提交的名字调用一个 bean，在 Tomcat 服务器创建一个 Web 服务目录。
6. 编写 2 个 JSP 页面 input.jsp 和 read.jsp。input.jsp 通过表单提交一个目录和该目录下的一个文件名给 read.jsp，read.jsp 根据 input.jsp 提交的目录和文件名调用一个 bean 读取文件的内容。

第 7 章

在 JSP 中使用数据库

> **本章导读**
> ✡ 知识点：掌握怎样在 JSP 中使用 JDBC 实现数据库的连接、查询、修改等操作。
> ✡ 重点：掌握连接数据库的常用方式——JDBC-ODBC 桥接器方式和加载 Java 数据库驱动程序方式。
> ✡ 难点：怎样实现数据库表的分页查询。
> ✡ 关键实践：编写 JSP 页面，使用 JavaBean 实现数据库表的分页查询。

Web 应用程序最常用的数据管理方式就是使用数据库，数据库已经成为 Web 程序开发的一个重要组成部分。本章将学习在 JSP 中怎样使用数据库。

本章大部分例子采用的模式是 JSP+JavaBean 模式，即 JSP 页面调用 bean 来完成对数据库的操作，如图 7.1 所示。为了便于教学，我们使用的数据库管理系统是 SQL Server 2000。在 JSP 中，只要掌握与某种数据库管理系统所管理的数据库交互信息，就会较容易地掌握与其他数据库管理系统所管理的数据库交互信息。本书将在 7.8 节介绍 JSP 怎样与其他常用的数据库管理系统交互信息。

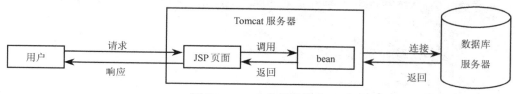

图 7.1 JSP 中使用数据库

本章使用的 Web 服务目录是 chapter7，为了使用 bean，在当前 Web 服务目录 chapter7 下建立目录 WEB-INF\classes。然后根据类的包名，在文件夹 classes 下建立相应的子目录。为了让 Tomcat 服务器启用上述目录，必须重新启动 Tomcat 服务器。

本章涉及数据库查询，为了方便解决中文乱码问题，我们采用的方案是将 page 指令中设置格式中的 charset 的首写字母小写：

```
<%@ page contentType="text/html;charset=GB2312" %>
```

7.1 SQL Server 2000 数据库管理系统

SQL Server 2000 数据库管理系统可以有效地管理数据库。SQL Server 2000 是一个网络数据库管理系统，也简称为数据库服务器。

1. 启动 SQL Server 2000

为了能让 Tomcat 服务器上的 Web 应用程序访问 SQL Server 2000 管理的数据库，必须启动 SQL Server 2000 提供的数据库服务器，即使 Tomcat 服务器和 SQL Server 2000 在同一计算机上也是如此。

如果已经安装 SQL Server 2000，可以用如下方式启动 SQL Server 2000 提供的数据库服务器："开始"→"程序"→"Microsoft SQL Server"→"服务器管理器"，出现如图 7.2 所示的窗口。

2. 建立数据库

打开"企业管理器"，出现如图所示 7.3 界面。"数据库"目录下是已有的数据库的名称，右键单击"数据库"，可以建立新的数据库，我们新建立的数据库名称是"Student"。

图 7.2　启动 SQL 2000 数据库服务器

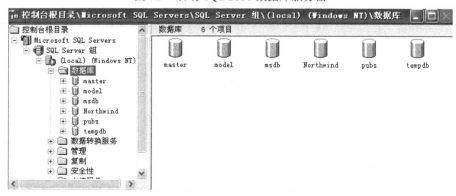

图 7.3　使用企业管理器建立新数据库

3. 创建表

在我们建立的数据库 Student 中创建名字为 score 的表。打开"企业管理器",选择"数据库"下的 Student,在 Student 管理的"表"的选项上单击右键,创建新的用户表:score。

score 表的字段(属性)为:

　　学号(char)　　姓名(char)　　数学成绩(float)　　物理成绩(float)　　英语成绩(float)

其中,"学号"字段为主键,如图 7.4 和图 7.5 所示。

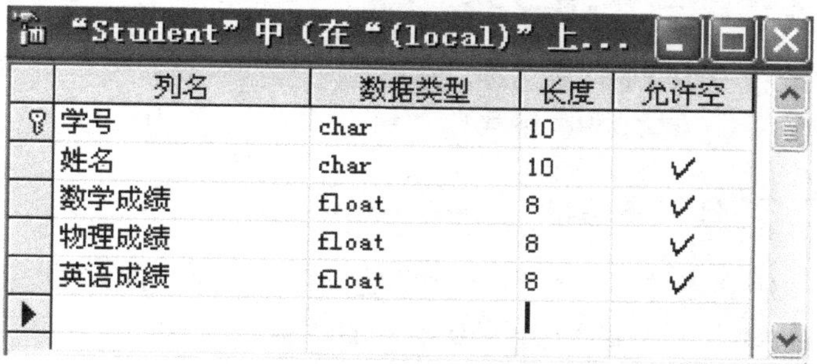

图 7.4　score 表及字段属性

图 7.5　在数据库 Student 中建立 score 表

7.2　JDBC

JDBC(Java DataBase Connectivity)是 Java 数据库连接 API,它由一些 Java 类和接口组成。在 JSP 中可以使用 JDBC 实现对数据库中表记录的查询、修改和删除等操作。JDBC 技术在 JSP 开发中占有很重要的地位。我们经常使用 JDBC 进行如下操作:

- 与一个数据库建立连接。
- 向已连接的数据库发送 SQL 语句。
- 处理 SQL 语句返回的结果。

使用 JDBC 时不需要知道底层数据库的细节,JDBC 操作不同的数据库仅仅是连接方式上的差异而已。使用 JDBC 的应用程序一旦与数据库建立连接,就可以使用 JDBC 提供的 API 操作数据库,如图 7.6 所示。

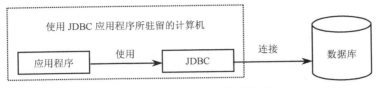

图 7.6 使用 JDBC 操作数据库

7.3 数据库连接的常用方式

应用程序必须首先与数据库建立连接，这也是与数据库交互信息所需的。本节介绍常用的两种连接方式：建立 JDBC-ODBC 桥接器和加载纯 Java 数据库驱动程序。

7.3.1 JDBC-ODBC 桥接器

JDBC 和数据库建立连接的一种常见方式是建立起一个 JDBC-ODBC 桥接器。ODBC（Open DataBase Connectivity）是 Microsoft 引进的数据库连接技术，主要目的是提供数据库访问的通用平台，由于 ODBC 驱动程序被广泛的使用，建立这种桥接器后，使得 JDBC 有能力访问几乎所有类型的数据库。

如果使用 JDBC-ODBC 桥接器访问数据库，事先必须设置数据源，因为 ODBC 通过数据源来管理数据，所以必须将某个数据库设置成 ODBC 的一个数据源，如图 7.7 所示。

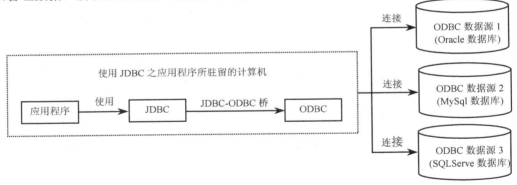

图 7.7 ODBC 的数据源

使用 JDBC-ODBC 桥接器方式与数据库建立连接，需要经过 3 个步骤：
<1> 创建 ODBC 数据源。
<2> 建立 JDBC-ODBC 桥接器。
<3> 与 ODBC 数据源指定的数据库建立连接。

假设应用程序所在的计算机要访问本地或远程的 SQL Server 2000 服务器上数据库，应用程序所在的计算机负责设置数据源，即将本地或远程的 SQL Server 服务器上的数据库设置成自己要访问的数据源。

1．创建 ODBC 数据源

创建 ODBC 数据源的步骤如下：
<1> 添加、修改或删除数据源。

选择"控制面板"→"管理工具"→"ODBC 数据源"（某些 Windows 系统需选择"控制面板"→"性能和维护"→"管理工具"→"ODBC 数据源"），然后双击 ODBC 数据源图标，出现如图 7.8 所示界面，该界面显示了用户已有的数据源的名称。选择"用户 DSN"，单击【添加】按钮，可以添加新的数据源；单击【配置】按钮，可以重新配置已有的数据源；单击【删除】按钮，可以删除已有的数据源。

<2> 为数据源选择驱动程序。

在图 7.8 所示的界面上选择单击【添加】按钮，出现为新增的数据源选择驱动程序界面（如图 7.9 所示），因为要访问 SQL Server 数据库，选择 SQL Server，然后单击【完成】按钮。

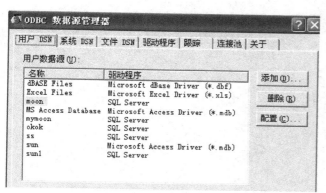
图 7.8 添加、修改或删除数据源

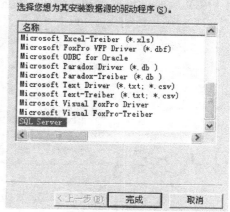
图 7.9 为新增的数据源选择驱动程序

<3> 数据源名称及所在位置。

在图 7.9 所示界面中单击【完成】按钮，将出现设置数据源具体项目的对话框（如图 7.10 所示）。在"名称"栏中为数据源起一个自己喜欢的名字，这里的名字是 mymoon（当然，如果你喜欢，可以把名字设为 mysun）。这个数据源就是指若干个数据库。在"你想连接哪个 SQL Server？"栏中选择或输入一个数据库服务器，这里选择网络上的一台机器：GXY（GXY 是本书 7.1 节中介绍的 SQL 2000 数据库服务器）。然后单击【下一步】按钮。

图 7.10 设置数据源的名字和所在服务器

<4> 设置 ID 与密码。

在图 7.10 所示界面中单击【下一步】按钮，出现设置 SQL Server 的 ID 和密码的界面（如图 7.11 所示），选择"使用用户输入登录标识号和密码的 SQL Server 验证"选项，这里设置用户名为"sa"、密码为"sa"（安装 SQL Server 2000 数据库服务器时设置的用户和密码）。然后单击【下一步】按钮。

<5> 选择数据库。

在出现的如图 7.12 所示界面中，选中"更改默认的数据库为"复选框，在其下拉菜单中选择用户 sa 有权限操作的 Student 数据库。当然，用户 sa 还可能有权操作其他数据库，这取决于 SQL server 2000 的有关设置，这里不再赘述。然后单击【下一步】按钮。

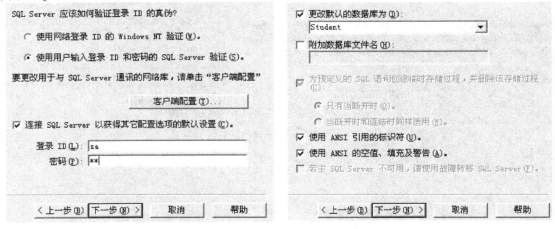

图 7.11　选择连接 SQL Server 的 ID 与密码　　　　图 7.12　选择数据库

<6> 创建数据源。

在出现的如图 7.13 所示界面中单击【完成】按钮，就创建了一个新的数据源：mymoon。

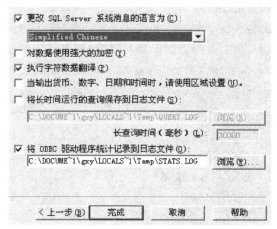

图 7.13　完成数据源设置

2．建立 JDBC-ODBC 桥接器

现在我们有了一个数据源 mymoon，这个数据源就是一个数据库——Student。为了连接这个数据库，首先要建立一个 JDBC-ODBC 桥接器：

```
Class.forName("sun.jdbc.odbc.JdbcOdbcDriver");
```
这里，Class 是包 java.lang 中的一个类，通过调用它的静态方法 forName 加载 sun.jdbc.odbc 包中的 JdbcOdbcDriver 类来建立 JDBC-ODBC 桥接器。

建立桥接器时可能发生异常，因此捕获这个异常。所以建立桥接器的标准如下：
```
try {
        Class.forName("sun.jdbc.odbc.JdbcOdbcDriver");
}
catch(ClassNotFoundException e) { }
```

3．与 ODBC 数据源指定的数据库建立连接

编写连接数据库代码不会出现数据库的名称，只能出现数据源的名字。先使用 java.sql 包中的 Connection 类声明一个对象，然后使用类 DriverManager 的静态方法 getConnection() 创建这个连接对象：
```
Connection con=DriverManager.getConnection("jdbc:odbc:数据源名字",
                                           "login name"," password ");
```
假如没有为数据源设置 login name 和 password，那么连接形式是：
```
Connection con=DriverManager. getConnection("jdbc:odbc:数据源名字","","");
```
为了能与数据源 mymoon 交换数据，建立 Connection 对象如下：
```
Connection con=DriverManager.getConnection("jdbc:odbc:mymoon", "sa","sa");
```
建立连接时应捕获 SQLException 异常：
```
try{
        Connection con=DriverManager.getConnection("jdbc:odbc:mymoon", "sa","sa");
}
catch(SQLException e) { }
```
这样就与数据源 mymoon 建立了连接。应用程序一旦与某个数据源建立连接，就可以通过 SQL 语句与该数据源所指定的数据库中的表交互信息，如查询、修改、更新表中的记录。

【例 7-1】制作一个简单的 JSP 页面，该页面中的 Java 程序片代码负责连接到数据源 mymoon，查询 score 表中的全部记录。效果如图 7.14 所示。

学号	姓名	数学成绩	英语成绩	物理成绩
0001	张三	90.0	89.0	78.0
0002	李四	67.0	78.0	90.0
0004	赵小林	89.0	79.0	89.0
0005	王近小	88.0	77.0	66.0

图 7.14 使用 JDBC-ODBC 桥接器连接数据库

showByJdbcOdbc.jsp
```
<%@ page contentType="text/html;Charset=GB2312" %>
```

```jsp
<%@ page import="java.sql.*" %>
<HTML><BODY>
  <% Connection con;
     Statement sql;
     ResultSet rs;
     try{ Class.forName("sun.jdbc.odbc.JdbcOdbcDriver"); }
     catch(ClassNotFoundException e) { out.print(e);  }
     try {
         con=DriverManager.getConnection("jdbc:odbc:mymoon","sa","sa");
         sql=con.createStatement();
         rs=sql.executeQuery("SELECT * FROM score");
         out.print("<table border=2>");
         out.print("<tr>");
         out.print("<th width=100>"+"学号");
         out.print("<th width=100>"+"姓名");
         out.print("<th width=50>"+"数学成绩");
         out.print("<th width=50>"+"英语成绩");
         out.print("<th width=50>"+"物理成绩");
         out.print("</TR>");
         while(rs.next()) {
             out.print("<tr>");
             out.print("<td >"+rs.getString(1)+"</td>");
             out.print("<td >"+rs.getString(2)+"</td>");
             out.print("<td >"+rs.getFloat("数学成绩")+"</td>");
             out.print("<td >"+rs.getFloat("物理成绩")+"</td>");
             out.print("<td >"+rs.getFloat("英语成绩")+"</td>");
             out.print("</tr>") ;
         }
         out.print("</table>");
         con.close();
     }
     catch(SQLException e1) { out.print(e1); }
  %>
</BODY></HTML>
```

7.3.2 使用纯 Java 数据库驱动程序

用 Java 语言编写的驱动程序称为纯 Java 驱动程序。JDBC 提供的 API 通过将纯 Java 驱动程序转换为 DBMS（数据库管理系统）所使用的专用协议来实现和特定的 DBMS 交互信息，简单地说，JDBC 可以调用本地的纯 Java 驱动程序和相应的数据库建立连接，如图 7.15 所示。

使用纯 Java 数据库驱动程序方式和数据库建立连接需要经过两个步骤：加载纯 Java 驱动程序 → 指定的数据库建立连接。

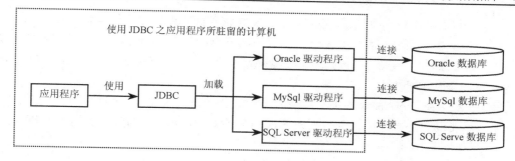

图 7.15 使用 Java 驱动程序

1. 加载纯 Java 驱动程序

与 JDBC-ODBC 桥方式不同的是，使用纯 Java 驱动程序访问数据库不需要设置数据源，由于不依赖于 ODBC，使得应用程序具有很好的移植性。目前，许多数据库厂商都提供了自己的相应的纯 Java 驱动程序。当使用纯 Java 驱动程序访问数据库时，必须要保证在连接数据库的应用程序所驻留的计算机上安装相应 DBMS 提供的纯 Java 驱动程序。比如，Tomcat 服务器上的某个 Web 应用程序想访问 SQL Server 2000 数据库管理系统所管理的数据库，Tomcat 服务器所驻留的计算机上必须要安装 SQL Server 2000 提供的纯 Java 驱动程序。

在微软官方网站可以下载 JDBC-ODBC 的驱动程序 sqljdbc_1.1.1501.101_enu.exe。安装 sqljdbc_1.1.1501.101_enu.exe 后，在安装目录的 enu 子目录中可以找到驱动程序文件 sqljdbc.jar，将该驱动程序复制到 Tomcat 服务器所使用的 JDK 的文件夹 \jre\lib\ext 中，如 D:\jdk1.5\jre\lib\ext，或复制到 Tomcat 服务器安装目录的文件夹 \common\lib 中，如 D:\apache-tomcat-5.5.20\common\lib。

应用程序加载 SQLServer 驱动程序代码如下：

```
try{
        Class.forName("com.microsoft.jdbc.sqlserver.SQLServerDriver").newInstance();
}
catch(Exception e) { }
```

2. 与指定的数据库建立连接

假设 SQL Server 数据库服务器所驻留的计算机的 IP 地址是 192.168.100.1，SQL Server 数据库服务器占用的端口是 1433。应用程序要与 SQL Server 数据库服务器管理的数据库 Student 建立连接，而有权访问数据库 Student 的用户的 id 和密码分别是 sa、sa，那么建立连接的代码如下：

```
try{
        String uri= "jdbc:sqlserver://192.168.100.1:1433;DatabaseName=Student";
        String user="sa";
        String password="sa";
        con=DriverManager.getConnection(uri,user,password);
}
catch(SQLException e) { }
```

应用程序一旦与某个数据库建立连接，就可以通过 SQL 语句与该数据库中的表交互信息，如查询、修改、更新表中的记录。

注：① 如果应用程序要连接 SQL Server 2000 服务器驻留在同一计算机上，使用的 IP 地址可以是 127.0.0.1；② 如果应用程序无法与 SQL Server 2000 服务器建立连接，可能需要更新 SQL Server 2000 服务器，登录到 http://www.microsoft.com 下载 SQL Server 2000 的补丁 SQLsp4.rar，安装该补丁即可。

【例 7-2】 制作一个简单的 JSP 页面，该页面中的 Java 程序片代码负责加载 SQL Server 2000 的 Java 驱动程序，并连接到数据库 Student，查询 score 表中的数学成绩大于 80 的全部记录。效果如图 7.16 所示。

学号	姓名	数学成绩	英语成绩	物理成绩
0001	张三	90.0	89.0	78.0
0004	赵小林	89.0	79.0	89.0
0005	王近小	88.0	77.0	66.0

图 7.16 使用纯 Java 数据库驱动程序连接数据库

showBySQLDriver.jsp

```jsp
<%@ page contentType="text/html;charset=GB2312" %>
<%@ page import="java.sql.*" %>
<HTML><BODY>
  <% Connection con;
     Statement sql;
     ResultSet rs;
     try {
         Class.forName("com.microsoft.sqlserver.jdbc.SQLServerDriver").newInstance();
     }
     catch(Exception e) { out.print(e); }
     try {
         String uri= "jdbc:sqlserver://127.0.0.1:1433;DatabaseName=Student";
         String user="sa";
         String password="sa";
         con=DriverManager.getConnection(uri,user,password);
         sql=con.createStatement();
         rs=sql.executeQuery("SELECT * FROM score WHERE  数学成绩 > 80");
         out.print("<table border=2>");
         out.print("<tr>");
         out.print("<th width=100>"+"学号");
         out.print("<th width=100>"+"姓名");
         out.print("<th width=50>"+"数学成绩");
         out.print("<th width=50>"+"英语成绩");
         out.print("<th width=50>"+"物理成绩");
         out.print("</TR>");
         while(rs.next()) {
             out.print("<tr>");
             out.print("<td >"+rs.getString(1)+"</td>");
```

```
                out.print("<td >"+rs.getString(2)+"</td>");
                out.print("<td >"+rs.getFloat("数学成绩")+"</td>");
                out.print("<td >"+rs.getFloat("物理成绩")+"</td>");
                out.print("<td >"+rs.getFloat("英语成绩")+"</td>");
                out.print("</tr>") ;
            }
            out.print("</table>");
            con.close();
        }
        catch(SQLException e1) { out.print(e1); }
    %>
</BODY></HTML>
```

7.4 查询操作

与数据库建立连接后,就可以使用 JDBC 提供的 API 和数据库交互信息,如查询、修改和更新数据库中的表等。JDBC 与数据库表进行交互的主要方式是使用 SQL 语句,JDBC 提供的 API 可以将标准的 SQL 语句发送给数据库,实现与数据库的交互。本书中后续的例子采用的数据库连接方式为 JDBC-ODBC 桥接器方式和使用纯 Java 驱动程序方式。使用的数据源的名字为 mymoon,数据库为 Student。

进行查询操作的具体步骤如下:

<1>向数据库发送 SQL 查询语句。

首先使用 Statement 声明一个 SQL 语句对象,然后利用 7.1 节创建的连接对象 con 调用方法 createStatment()创建这个 SQL 语句对象。

```
        try {
                Statement  sql=con.createStatement();
        }
        catch(SQLException e ) { }
```

<2>处理查询结果。

有了 SQL 语句对象后,这个对象就可以调用相应的方法,实现对数据库中表的查询和修改,并将查询结果存放在一个 ResultSet 类声明的对象中。也就是说,SQL 查询语句对数据库的查询操作将返回一个 ResultSet 对象。ResultSet 对象是以统一形式的列组织的数据行组成。

例如,对于

```
        ResultSet   rs=sql.executeQuery("SELECT *   FROM score");
```
内存的结果集对象 rs 如图 7.17(a)所示;而对于
```
        ResultSet   rs=sql.executeQuery("SELECT 姓名,物理成绩,英语成绩   FROM score");
```
内存的结果集对象 rs 如图 7.17(b)所示。

ResultSet 对象一次只能看到一个数据行,使用 next()方法走到下一数据行,获得一行数据后,ResultSet 对象可以使用 getXxx 方法获得字段值,将位置索引(第一列使用 1,第二列使用 2,……)或字段名传递给 getXxx 方法的参数即可。表 7.1 给了出了 ResultSet 对象的若干方法。

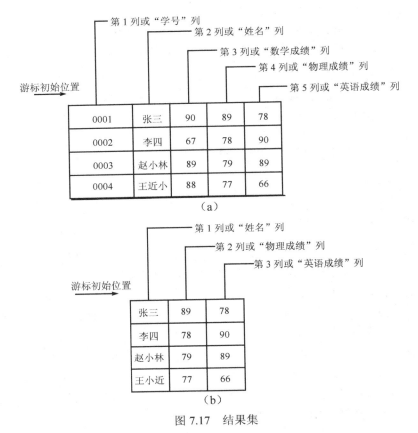

图 7.17 结果集

表 7.1 ResultSet 类的若干方法

方法名称	返回类型	方法名称	返回类型
next()	boolean	getByte(String columnName)	byte
getByte(int columnIndex)	byte	getDate(String columnName)	Date
getDate(int columnIndex)	Date	getDouble(String columnName)	double
getDouble(int columnIndex)	double	getFloat(String columnName)	float
getFloat(int columnIndex)	float	getInt(String columnName)	int
getInt(int columnIndex)	int	getLong(String columnName)	long
getLong(int columnIndex)	long	getString(String columnName)	String
getString(int columnIndex)	String		

无论字段是何种属性，总可以使用 getSring() 方法返回字段值的串表示；当使用 ResultSet 的 getXxx 方法查看一行记录时，不可以颠倒字段的顺序。例如：

    rs.getFloat(5);　　　　　　　　//错误
    rs.getFloat(4);　　　　　　　　//错误

7.4.1　顺序查询

查询数据库中的一个表的记录时，我们希望知道表中字段的个数以及各字段的名字。由于无论字段是何种属性,总可以使用 getSring()方法返回字段值的串表示。因此,只要知道了表中字段的个数或字段的名字,就可以方便地查询表中的记录。

那么,怎样知道一个表中有哪些字段呢?通过使用 JDBC 提供的 API,可以在查询之前知道表中的字段的个数和名字,这有助于编写可复用的查询代码。

与数据库建立连接之后,比如:

  con=DriverManager.getConnection("jdbc:odbc:mymoon","sa","sa");

那么,该连接对象调用 getMetaData()方法可以返回一个 DatabaseMetaData 对象,例如:

  DatabaseMetaData metadata=con.getMetaData();

metadata 对象再调用 getColumns 可以将 score 表的字段信息以行列的形式存储在一个 ResultSet 对象中,例如:

  ResultSet tableMessage=metadata.getColumns(null,null, "score",null);

如果 score 表有 n 个字段,tableMessage 就刚好有 n 行、每行 4 列。每行分别含有与相应字段有关的信息,信息的次序为:"数据库名"、"数据库扩展名"、"表名"、"字段名"。例如,score 表有 5 个字段,那么上述 tableMessage 刚好有 5 行,每行有 4 列,如图 7.18 所示。

Student	dbo	score	学习
Student	dbo	score	姓名
Student	dbo	score	数学成绩
Student	dbo	score	物理成绩
Student	dbo	score	英语成绩

图 7.18 tableMessage 对象

tableMessage 对象调用 next 方法使游标向下移动一行(游标的初始位置在第 1 行之前),然后 tableMessage 调用 getXxx()方法可以查看该行中列的信息,其中最重要的信息是第 4 列,该列上的信息为字段的名字。

【例 7-3】 编写两个 bean,其中一个使用 JDBC-ODBC 桥接器方式连接数据库、查询表,另一个使用加载纯 Java 数据库驱动程序方式连接数据库、查询表。用户可以通过 JSP 页面 main.jsp 提供的表单输入数据库(数据源)和表的名字。JSP 页面 inquireOne.jsp 和 inquireTwo.jsp 分别显示两个 bean 查询的有关结果。

创建 bean 的源文件如下:

QueryBeanOne.java

```
package tom.jiafei;
import java.sql.*;
public class QueryBeanOne {
    String ODBCDataSource="";           //ODBC 数据源名称
    String tableName="";                //表的名字
    String user=""         ;            //用户
    String secret="" ;                  //密码
    StringBuffer queryResult;           //查询结果
    public QueryBeanOne() {
        queryResult=new StringBuffer();
        try { Class.forName("sun.jdbc.odbc.JdbcOdbcDriver"); }
```

```java
            catch(ClassNotFoundException e) {
                queryResult=new StringBuffer();
                queryResult.append(""+e);
            }
    }
    public void setODBCDataSource(String s) {
        ODBCDataSource=s.trim();
    }
    public String getODBCDataSource() {
        return ODBCDataSource;
    }
    public void setTableName(String s) {
        tableName=s.trim();
    }
    public String getTableName(){
        return tableName;
    }
    public void setSecret(String s) {
        secret=s.trim();;
    }
    public String getSecret(){
        return secret;
    }
    public void setUser(String s) {
        user=s.trim();;
    }
    public String getUser(){
        return user;
    }
    public StringBuffer getQueryResult(){
        Connection con;
        Statement sql;
        ResultSet rs;
        try {  queryResult.append("<table border=1>");
            String source="jdbc:odbc:"+ODBCDataSource;
            String id=user;
            String password=secret;
            con=DriverManager.getConnection(source,id,password);
            DatabaseMetaData metadata=con.getMetaData();
            //如果 tableName 表有 n 个字段，rs1 就刚好有 n 行，每行中含有字段名的信息
            ResultSet rs1=metadata.getColumns(null,null,tableName,null);
            int 字段个数=0;
            queryResult.append("<tr>");
            while(rs1.next()) {            //
                字段个数++;
                String clumnName=rs1.getString(4);
                queryResult.append("<td>"+clumnName+"</td>");
```

```
                }
                queryResult.append("</tr>");
                sql=con.createStatement();
                rs=sql.executeQuery("SELECT * FROM "+tableName);
                while(rs.next()) {
                    queryResult.append("<tr>");
                    for(int k=1;k<=字段个数;k++) {
                        queryResult.append("<td>"+rs.getString(k)+"</td>");
                    }
                    queryResult.append("</tr>");
                }
                queryResult.append("</table>");
                con.close();
            }
            catch(SQLException e) { queryResult.append(e); }
            return queryResult;
        }
    }
```

QueryBeanTwo.java

```
    package tom.jiafei;
    import java.sql.*;
    public class QueryBeanTwo{
        String databaseName="";              //数据库名称
        String tableName="";                 //表的名字
        String user=""         ;             //用户
        String secret="" ;                   //密码
        StringBuffer queryResult;            //查询结果
        public QueryBeanTwo() {
            queryResult=new StringBuffer();
            try{ Class.forName("com.microsoft.sqlserver.jdbc.SQLServerDriver").newInstance(); }
            catch(Exception e) {
                queryResult=new StringBuffer();
                queryResult.append(""+e);
            }
        }
        public void setDatabaseName(String s) {
            databaseName=s.trim();
        }
        public String getDatabaseName(){
            return databaseName;
        }
        public void setTableName(String s) {
            tableName=s.trim();
        }
        public String getTableName(){
            return tableName;
        }
```

```java
    public void setSecret(String s) {
        secret=s.trim();;
}
    public String getSecret()
    {
        return secret;
    }
    public void setUser(String s) {
        user=s.trim();;
}
    public String getUser(){
        return user;
}
    public StringBuffer getQueryResult(){
        Connection con;
        Statement sql;
        ResultSet rs;
        try {   queryResult.append("<table border=1>");
            String uri= "jdbc:sqlserver://127.0.0.1:1433;DatabaseName="+databaseName;
            String id=user;
            String password=secret;
            con=DriverManager.getConnection(uri,id,password);
            DatabaseMetaData metadata=con.getMetaData();
            //如果 tableName 表有 n 个字段，rs1 就刚好有 n 行，每行中含有字段名的信息
            ResultSet rs1=metadata.getColumns(null,null,tableName,null);
            int 字段个数=0;
            queryResult.append("<tr>");
            while(rs1.next()){
                字段个数++;
                String clumnName=rs1.getString(4);
                queryResult.append("<td>"+clumnName+"</td>");
            }
            queryResult.append("</tr>");
            sql=con.createStatement();
            rs=sql.executeQuery("SELECT * FROM "+tableName);
            while(rs.next()){
                queryResult.append("<tr>");
                for(int k=1;k<=字段个数;k++) {
                    queryResult.append("<td>"+rs.getString(k)+"</td>");
                }
                queryResult.append("</tr>");
            }
            queryResult.append("</table>");
            con.close();
        }
        catch(SQLException e) { queryResult.append(e); }
        return queryResult;
```

```
        }
    }
```

JSP 页面：

main.jsp（效果如图 7.19 所示）

```
<%@ page contentType="text/html;charset=GB2312" %>
<HTML><BODY ><FONT size=2>
<H5>使用 ODBC 数据源连接、查询数据库:</H5>
<FORM action="inquireOne.jsp" method="post" >
  <BR>输入 ODBC 数据源的名字:
  <Input type=text name="ODBCDataSource" value="mymoon" size=8>
    输入表的名字:
  <Input type=text name="tableName" value="score" size=8>
  <BR>输入用户名称:
  <Input type=text name="user" size=6>
    输入密码:
  <Input type="password" name="secret" size=6>
  <Input type=submit name="g" value="提交">
</FORM>
<H5>使用纯 Java 驱动程序连接、查询数据库:</H5>
<FORM action="inquireTwo.jsp" Method="post" >
    输入数据库的名字
  <Input type=text name="databaseName" value="Student" size=8>
    输入表的名字
  <Input type=text name="tableName" value="score" size=8>
  <BR>输入用户名称:
  <Input type=text name="user" size=6>
    输入密码:
  <Input type="password" name="secret" size=6>
  <Input type=submit name="g" value="提交">
</FORM>
</BODY></HTML>
```

inquireOne.jsp（效果如图 7.20 所示）

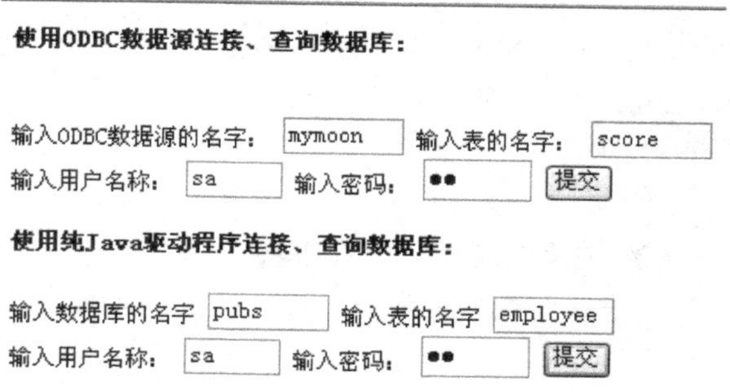

图 7.19　选择数据源或数据库

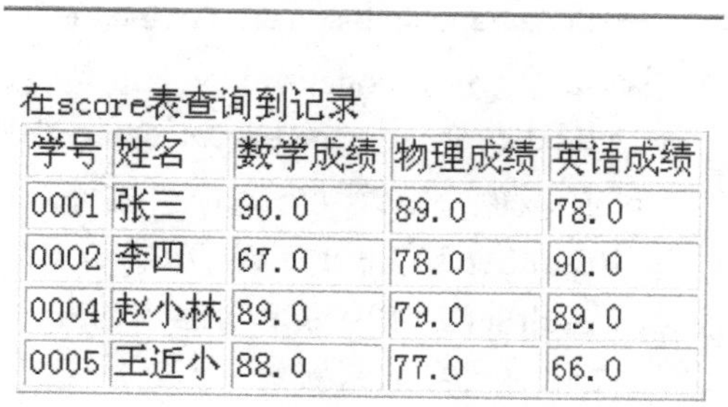

在score表查询到记录

学号	姓名	数学成绩	物理成绩	英语成绩
0001	张三	90.0	89.0	78.0
0002	李四	67.0	78.0	90.0
0004	赵小林	89.0	79.0	89.0
0005	王近小	88.0	77.0	66.0

图 7.20　使用 JDBC-ODBC 桥接器方式查询

```
<%@ page contentType="text/html;charset=GB2312" %>
<%@ page import="tom.jiafei.QueryBeanOne" %>
<jsp:useBean id="database" class="tom.jiafei.QueryBeanOne" scope="request" />
<jsp:setProperty   name= "database"   property="*" />
```

```
<HTML><BODY>
<BR>
    在<jsp:getProperty name= "database" property="tableName" />表查询到记录
<BR>
<jsp:getProperty name= "database" property="queryResult" />
</BODY></HTML>
```
inquireTwo.jsp（效果如图 7.21 所示）

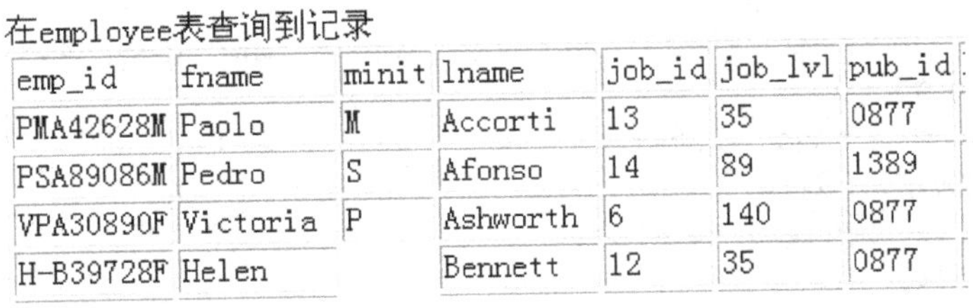

在employee表查询到记录

emp_id	fname	minit	lname	job_id	job_lvl	pub_id
PMA42628M	Paolo	M	Accorti	13	35	0877
PSA89086M	Pedro	S	Afonso	14	89	1389
VPA30890F	Victoria	P	Ashworth	6	140	0877
H-B39728F	Helen		Bennett	12	35	0877

图 7.21　使用纯 Java 驱动程序方式查询

```
<%@ page contentType="text/html;charset=GB2312" %>
<%@ page import="tom.jiafei.QueryBeanTwo" %>
<jsp:useBean id="database" class="tom.jiafei.QueryBeanTwo" scope="request" />
<jsp:setProperty   name= "database"   property="*" />
<BR>
    在<jsp:getProperty   name= "database"   property="tableName"   />表查询到记录
<BR>
<jsp:getProperty   name= "database"   property="queryResult"   />
</BODY>
</HTML>
```

7.4.2　随机查询

前面学习了使用 ResultSet 的 next()方法顺序查询数据，但有时需要在结果集中前后移动，显示结果集指定的一条记录，或随机显示若干条记录等，这时必须返回一个可滚动的结果集。

为了得到一个可滚动的结果集，需使用下述方法先获得一个 Statement 对象：

　　Statement stmt=con.createStatement(int type ,int concurrency);

然后根据参数 type、concurrency 的取值情况，stmt 返回相应类型的结果集：

　　ResultSet   re=stmt.executeQuery(SQL 语句);

type 的取值决定滚动方式，取值可以是：

- ResultSet.TYPE_FORWORD_ONLY ——结果集的游标只能向下滚动。
- ResultSet.TYPE_SCROLL_INSENSITIVE ——结果集的游标可以上下移动，当数据库变化时，当前结果集不变。
- ResultSet.TYPE_SCROLL_SENSITIVE ——返回可滚动的结果集，当数据库变化时，当前结果集同步改变。

Concurrency 的取值决定是否可以用结果集更新数据库，取值可以是：

- ResultSet.CONCUR_READ_ONLY ——不能用结果集更新数据库中的表。
- ResultSet.CONCUR_UPDATABLE ——能用结果集更新数据库中的表。

滚动查询经常用到 ResultSet 的下述方法：
- public boolean previous() ——将游标向上移动，该方法返回 boolean 型数据，当移到结果集第一行之前时返回 false。
- public void beforeFirst() ——将游标移动到结果集的初始位置，即在第一行之前。
- public void afterLast() ——将游标移到结果集最后一行之后。
- public void first() ——将游标移到结果集的第一行。
- public void last() ——将游标移到结果集的最后一行。
- public boolean isAfterLast() ——判断游标是否在最后一行之后。
- public boolean isBeforeFirst() ——判断游标是否在第一行之前。
- public boolean ifFirst() ——判断游标是否指向结果集的第一行。
- public boolean isLast() ——判断游标是否指向结果集的最后一行。
- public int getRow() ——得到当前游标所指行的行号，行号从 1 开始，如果结果集没有行，返回 0。
- public boolean absolute(int row) ——将游标移到参数 row 指定的行号。

注意：如果 row 取负值，就是倒数的行数，absolute(-1)表示移到最后一行，absolute(-2)表示移到倒数第 2 行。当移动到第一行前面或最后一行的后面时，该方法返回 false。

【例 7-4】 负责查询数据库的 bean 首先将游标移动到最后一行，再获取最后一行的行号，以便获得表中的记录数目，该 bean 能随机输出若干条记录。

创建 bean 的源文件如下：

RandomQueryBean.java

```java
    package tom.jiafei;
    import java.sql.*;
    import java.util.*;
    public class RandomQueryBean {
        String databaseName="";              //数据库名称
        String tableName="";                 //表的名字
        int count;                           //记录总数
        int randomNumber;                    //随机输出的记录数
        StringBuffer randomQueryResult;      //查询结果
        Connection con;
        Statement sql;
        ResultSet rs;
        public RandomQueryBean(){
            randomQueryResult=new StringBuffer();
            try{ Class.forName("com.microsoft.sqlserver.jdbc.SQLServerDriver").newInstance(); }
            catch(Exception e) {
                randomQueryResult=new StringBuffer();
                randomQueryResult.append(""+e);
            }
        }
        public void setDatabaseName(String s) {
```

```java
            databaseName=s.trim();
        }
        public String getDatabaseName(){
            return databaseName;
        }
        public void setTableName(String s) {
            tableName=s.trim();
        }
        public String getTableName(){
            return tableName;
        }
        public void setRandomNumber(int n) {
            randomNumber=n;
        }
        public int getRandomNumber(){
            return randomNumber;
        }
        public int getCount(){
            try{  String uri="jdbc:sqlserver://127.0.0.1:1433;DatabaseName="+databaseName;
                con=DriverManager.getConnection(uri,"sa","sa");
                sql= con.createStatement(ResultSet.TYPE_SCROLL_SENSITIVE,
                                        ResultSet.CONCUR_READ_ONLY);
                rs=sql.executeQuery("SELECT * FROM "+tableName);
                rs.last();
                count=rs.getRow();
                con.close();
            }
            catch(SQLException exp) { count=-1;   }
            return count;
        }
        public StringBuffer getRandomQueryResult(){
            randomQueryResult=new StringBuffer();
            try {   randomQueryResult.append("<table border=1>");
                String uri= "jdbc:sqlserver://127.0.0.1:1433;DatabaseName="+databaseName;
                con=DriverManager.getConnection(uri,"sa","sa");
                DatabaseMetaData metadata=con.getMetaData();
                //如果 tableName 表有 n 个字段, rs1 就刚好有 n 行, 每行中含有字段名的信息
                ResultSet rs1=metadata.getColumns(null,null,tableName,null);
                int  字段个数=0;
                randomQueryResult.append("<tr>");
                while(rs1.next()){
                    字段个数++;
                    String clumnName=rs1.getString(4);
```

```
                    randomQueryResult.append("<td>"+clumnName+"</td>");
                }
                randomQueryResult.append("</tr>");
                sql=con.createStatement(ResultSet.TYPE_SCROLL_SENSITIVE,
                                        ResultSet.CONCUR_READ_ONLY);
                rs=sql.executeQuery("SELECT * FROM "+tableName);
                rs.last();
                count=rs.getRow();
                Vector<Integer> vector=new Vector<Integer>();
                for(int i=1;i<=count;i++){
                    vector.add(new Integer(i));
                }
                int 抽取数目=Math.min(randomNumber,count);
                while(抽取数目>0) {
                    int i=(int)(Math.random()*vector.size());
                    int index=(vector.elementAt(i)).intValue();   //在向量 vector 中随机抽取一个元素
                    rs.absolute(index);                            //游标移到这一行
                    randomQueryResult.append("<tr>");
                    for(int k=1;k<=字段个数;k++) {
                        randomQueryResult.append("<td>"+rs.getString(k)+"</td>");
                    }
                    randomQueryResult.append("</tr>");
                    抽取数目--;
                    vector.removeElementAt(i);                     //将抽取过的元素从向量中删除
                }
                randomQueryResult.append("</table>");
                con.close();
            }
            catch(SQLException e) { randomQueryResult.append(e); }
            randomNumber=0;
            return randomQueryResult;
        }
    }
```

JSP 页面文件如下：

main.jsp（效果如图 7.22 所示）

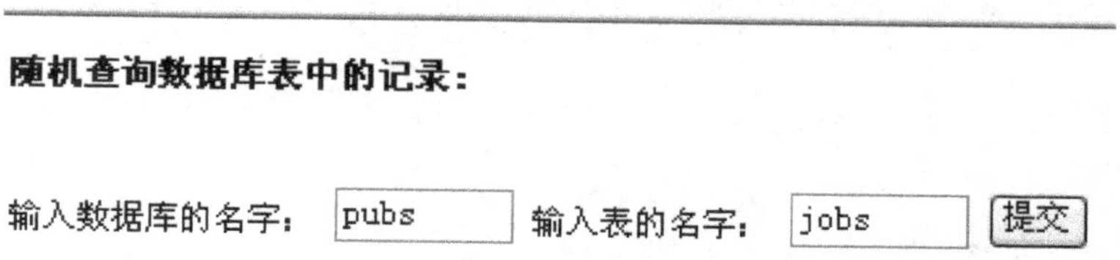

随机查询数据库表中的记录：

输入数据库的名字：　pubs　　输入表的名字：　jobs　　提交

图 7.22　选择数据库

```
<%@ page contentType="text/html;charset=GB2312" %>
<HTML><BODY ><FONT size=2>
```

```
<H5>随机查询数据库表中的记录:</H5>
<FORM action="randomQuery.jsp" method="post" >
  <BR>输入数据库的名字:
  <Input type=text name="databaseName" value="pubs" size=8>
   输入表的名字:
  <Input type=text name="tableName" value="jobs" size=8>
  <Input type=submit name="g" value="提交">
</FORM>
</BODY></HTML>
```

randomQuery.jsp（效果如图 7.23 所示）

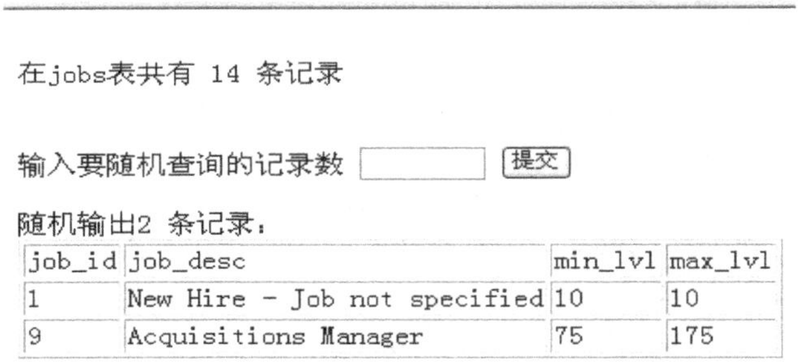

在jobs表共有 14 条记录

输入要随机查询的记录数 [提交]

随机输出2 条记录：

job_id	job_desc	min_lvl	max_lvl
1	New Hire - Job not specified	10	10
9	Acquisitions Manager	75	175

图 7.23　随机输出记录查询

```
<%@ page contentType="text/html;charset=GB2312" %>
<%@ page import="tom.jiafei.RandomQueryBean" %>
<jsp:useBean id="database" class="tom.jiafei.RandomQueryBean" scope="session" />
<jsp:setProperty  name= "database"  property="*" />
<BR>
   在<jsp:getProperty  name= "database"  property="tableName"  />表共有
   <jsp:getProperty  name= "database"  property="count"  />
  条记录
 <FORM action="" method="post" >
  <BR>输入要随机查询的记录数
  <Input type=text name="randomNumber"  size=8>
  <Input type=submit name="g" value="提交">
</FORM>
<jsp:setProperty  name= "database"  property="randomNumber" param="randomNumber" />
  随机输出<jsp:getProperty  name= "database"  property="randomNumber"  />
  条记录:
  <jsp:getProperty  name= "database"  property="randomQueryResult"  />
<BR>
</BODY>
</HTML>
```

7.4.3　条件查询

【例 7-5】 客户通过 JSP 页面输入查询条件，如按姓名查询成绩或按分数段查询 score 表中的记录等。该例子中的 bean 可以根据 JSP 页面提交的数据进行相应的数据库查询操作。

创建 bean 的源文件如下：

ConditionQuery.java

```java
package tom.jiafei;
import java.sql.*;
public class ConditionQuery {
    String number ;                                    //学号
    String personName;                                 //姓名
    float mathMax,mathMin;                             //查询的数学成绩的最高和最低分
    StringBuffer queryResultByName,                    //查询结果
                 queryResultByNumber,
                 queryResultByScore;
    public ConditionQuery() {
       queryResultByName=new StringBuffer();
       queryResultByNumber=new StringBuffer();
       queryResultByScore=new StringBuffer();
       try { Class.forName("sun.jdbc.odbc.JdbcOdbcDriver"); }
       catch(ClassNotFoundException e) { }
    }
    public void setNumber(String s) {
       number=s.trim();
    }
    public String getNumber(){
       return number;
    }
    public void setPersonName(String s) {
       personName=s.trim();
       try{   byte bb[]=personName.getBytes("ISO-8859-1");
            personName=new String(bb,"gb2312");
          }
       catch(Exception e) { }
    }
    public String getPersonName(){
       return personName;
    }
    public void setMathMax(float n) {
       mathMax=n;
    }
    public float getMathMax(){
       return mathMax;
    }
    public void setMathMin(float n) {
       mathMin=n;
    }
```

```java
    public float getMathMin(){
        return mathMin;
    }
    public StringBuffer getQueryResultByName(){
        String condition="SELECT * FROM score Where 姓名 = '"+personName+"'";
        queryResultByName=f(condition);
        return queryResultByName;
    }
    public StringBuffer getQueryResultByNumber(){
        String condition="SELECT * FROM score Where 学号 = '"+number+"'";
        queryResultByNumber=f(condition);
        return queryResultByNumber;
    }
    public StringBuffer getQueryResultByScore(){
        String condition="SELECT * FROM score Where 数学成绩 <= "+mathMax
                                    +" AND "+"数学成绩 >= "+mathMin;
        queryResultByScore=f(condition);
        return queryResultByScore;
    }
    private StringBuffer f(String condition) {
        StringBuffer str=new StringBuffer();
        Connection con;
        Statement sql;
        ResultSet rs;
        try {  con=DriverManager.getConnection("jdbc:odbc:mymoon","sa","sa");
            sql=con.createStatement();
            rs=sql.executeQuery(condition);
            str.append("<table border=1>");
            str.append("<th width=100>"+"学号");
            str.append("<th width=100>"+"姓名");
            str.append("<th width=100>"+"数学成绩");
            str.append("<th width=100>"+"物理成绩");
            str.append("<th width=100>"+"英语成绩");
            while(rs.next()){
                str.append("<tr>");
                str.append("<td>"+rs.getString(1)+"</td>");
                str.append("<td>"+rs.getString(2)+"</td>");
                str.append("<td>"+rs.getString(3)+"</td>");
                str.append("<td>"+rs.getString(4)+"</td>");
                str.append("<td>"+rs.getString(5)+"</td>");
                str.append("</tr>");
            }
            str.append("<table border=1>");
```

```
                con.close();
            }
            catch(SQLException e) { str.append(e); }
            return str;
        }
    }
```

JSP 页面文件如下：

inputCondition.jsp（效果如图 7.24 所示）

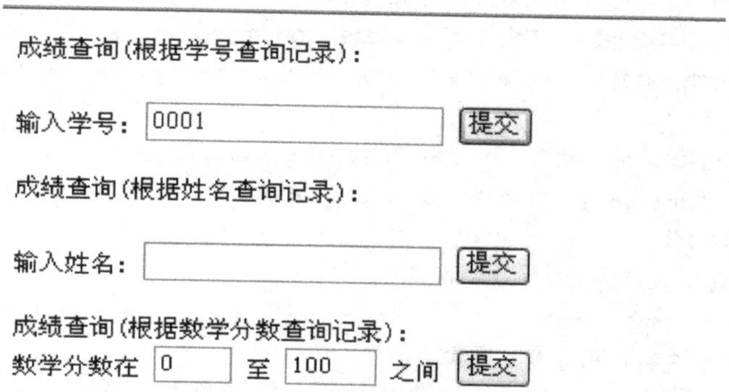

图 7.24 输入查询条件

```
<%@ page contentType="text/html;charset=GB2312" %>
<HTML><BODY>
<FONT size=2>
<FORM action="inquireOne.jsp" Method="post">
   <P>成绩查询(根据学号查询记录):
   <P>输入学号:
   <Input type=text name="number">
   <Input type=submit name="g" value="提交">
</FORM>
<FORM action="inquireTwo.jsp" Method="post">
   <P>成绩查询(根据姓名查询记录):
   <P>输入姓名:
   <Input type=text name="personName">
   <Input type=submit name="g" value="提交">
</FORM>
<FORM action="inquireThree.jsp" Method="post" >
   <P>成绩查询(根据数学分数查询记录):
   <BR> 数学分数在
      <Input type=text name="mathMin" value=0 size=4>
      至
      <Input type=text name="mathMax" value=100 size=5>
      之间
      <Input type=submit   value="提交">
</FORM>
</BODY></HTML>
```

inquireOne.jsp（效果如图 7.25 所示）

根据学号0001查询到记录				
学号	姓名	数学成绩	物理成绩	英语成绩
0001	张三	90.0	89.0	78.0

图 7.25 根据学号查询

```
<%@ page contentType="text/html;charset=GB2312" %>
<%@ page import="tom.jiafei.ConditionQuery" %>
<jsp:useBean id="database" class="tom.jiafei.ConditionQuery" scope="request" />
<jsp:setProperty   name= "database"   property="number" param="number"/>
<HTML><BODY>
   根据学号<jsp:getProperty  name= "database"   property="number"   />查询到记录
<BR>  <jsp:getProperty  name= "database"   property="queryResultByNumber" />
</BODY></HTML>
```

inquireTwo.jsp（效果如图 7.26 所示）

根据姓名李四查询到记录				
学号	姓名	数学成绩	物理成绩	英语成绩
0002	李四	67.0	78.0	90.0

图 7.26 根据姓名查询

```
<%@ page contentType="text/html;charset=GB2312" %>
<%@ page import="tom.jiafei.ConditionQuery" %>
<jsp:useBean id="database" class="tom.jiafei.ConditionQuery" scope="request" />
<jsp:setProperty   name= "database"   property="personName" param="personName"/>
<HTML><BODY>
   根据姓名<jsp:getProperty  name= "database"   property="personName"   />查询到记录
<BR>  <jsp:getProperty  name= "database"   property="queryResultByName" />
</BODY></HTML>
```

inquireThree.jsp（效果如图 7.27 所示）

根据数学分数在85.0至 89.0 之间查询到的记录				
学号	姓名	数学成绩	物理成绩	英语成绩
0004	赵小林	89.0	79.0	89.0
0005	王近小	88.0	77.0	66.0

图 7.27 根据成绩查询

```
<%@ page contentType="text/html;charset=GB2312" %>
<%@ page import="tom.jiafei.ConditionQuery" %>
<jsp:useBean id="database" class="tom.jiafei.ConditionQuery" scope="request" />
<jsp:setProperty   name= "database"   property="mathMin" param="mathMin"/>
<jsp:setProperty   name= "database"   property="mathMax" param="mathMax"/>
<HTML><BODY>
   根据数学分数在<jsp:getProperty  name= "database"   property="mathMin"   />至
```

　　　　　　　　　<jsp:getProperty name= "database" property="mathMax" />
　　之间查询到的记录
　　
　<jsp:getProperty name= "database" property="queryResultByScore" />
　　</BODY></HTML>

7.4.4 排序查询

可以在 SQL 语句中使用 ORDER BY 子语句，对记录排序。例如，按总成绩排序查询的 SQL 语句如下：

　　SELECT * FROM score　ORDER BY 数学成绩+物理成绩+英语成绩

【例 7-6】 通过 JSP 页面可以选择按 3 科的总分从低到高排列记录、按姓氏拼音排序或英语成绩排序记录。bean 根据 JSP 页面的选择的排序方式，使用 SQL 语句的 ORDER BY 子语句排序记录。

创建 bean 的源文件如下：

OrderQuery.java

```java
package tom.jiafei;
import java.sql.*;
public class OrderQuery {
    String orderType;            //排序方式
    StringBuffer orderResult;    //排序结果
    public OrderQuery(){
        orderResult=new StringBuffer();
        try { Class.forName("sun.jdbc.odbc.JdbcOdbcDriver"); }
        catch(ClassNotFoundException e) { }
    }
    public void setOrderType(String s) {
        orderType=s.trim();
        try{  byte bb[]=orderType.getBytes("ISO-8859-1");
              orderType=new String(bb,"gb2312");
        }
        catch(Exception e){ }
    }
    public String getOrderType(){
        return orderType;
    }
    public StringBuffer getOrderResult(){
        String orderCondition="SELECT * FROM score ORDER BY "+orderType;
        Connection con;
        Statement sql;
        ResultSet rs;
        try { con=DriverManager.getConnection("jdbc:odbc:mymoon","sa","sa");
            sql=con.createStatement();
            rs=sql.executeQuery(orderCondition);
```

```
            orderResult.append("<table border=1>");
            orderResult.append("<th width=80>"+"学号");
            orderResult.append("<th width=80>"+"姓名");
            orderResult.append("<th width=80>"+"数学成绩");
            orderResult.append("<th width=80>"+"物理成绩");
            orderResult.append("<th width=80>"+"英语成绩");
            orderResult.append("<th width=80>"+"总成绩");
            while(rs.next()){
                orderResult.append("<tr>");
                orderResult.append("<td>"+rs.getString(1)+"</td>");
                orderResult.append("<td>"+rs.getString(2)+"</td>");
                float mathScore=rs.getFloat(3);
                orderResult.append("<td>"+mathScore+"</td>");
                float phsicsScore=rs.getFloat(4);
                orderResult.append("<td>"+phsicsScore+"</td>");
                float englishScore=rs.getFloat(5);
                orderResult.append("<td>"+englishScore+"</td>");
                float sum=mathScore+phsicsScore+englishScore;
                orderResult.append("<th>"+sum+"</th>");
                orderResult.append("</tr>");
            }
            orderResult.append("<table border=1>");
            con.close();
        }
        catch(SQLException e) { orderResult.append(e); }
        return orderResult;
    }
}
```

JSP 页面文件如下：

choiceType.jsp（效果如图 7.28 所示）

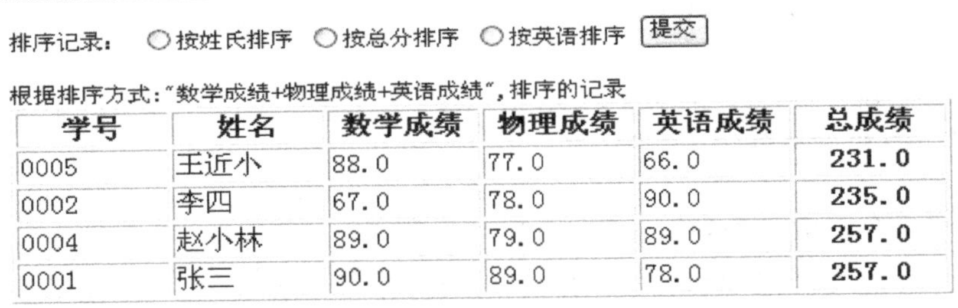

学号	姓名	数学成绩	物理成绩	英语成绩	总成绩
0005	王近小	88.0	77.0	66.0	231.0
0002	李四	67.0	78.0	90.0	235.0
0004	赵小林	89.0	79.0	89.0	257.0
0001	张三	90.0	89.0	78.0	257.0

图 7.28　排序记录

```
<%@ page contentType="text/html;charset=GB2312" %>
<%@ page import="tom.jiafei.OrderQuery" %>
<jsp:useBean id="database" class="tom.jiafei.OrderQuery" scope="request" />
<jsp:setProperty   name= "database"   property="orderType" param="orderType"/>
```

```html
<HTML><BODY><FONT size=2>
<FORM action="" method=post name=form>
   排序记录：
   <Input type="radio" name="orderType" value="姓名">按姓氏排序
   <Input type="radio" name="orderType" value="数学成绩+物理成绩+英语成绩">按总分排序
   <Input type="radio" name="orderType" value="英语成绩">按英语排序
   <Input type=submit name="g" value="提交">
</FORM>
根据排序方式:"<jsp:getProperty name= "database"  property="orderType"  />",排序的记录
<BR> <jsp:getProperty   name= "database"   property="orderResult" />
</BODY></HTML>
```

7.4.5 模糊查询

可以用 SQL 语句操作符 LIKE 进行模式般配，使用 "%" 代替一个或多个字符，用一个下划线 "_" 代替一个字符。比如，下述语句查询姓氏是 "王" 的记录：

```
rs=sql.executeQuery("SELECT * FROM students WHERE 姓名 LIKE '王%' ");
```

【例 7-7】 JSP 页面负责选择模糊查询条件，bean 负责连接数据库查询记录。
创建 bean 的源文件：

KeyWordQuery.java

```java
package tom.jiafei;
import java.sql.*;
public class KeyWordQuery {
    String ziduan="",                          //字段
           keyWord="";                         //关键字
    StringBuffer queryResult;
    public KeyWordQuery(){
        queryResult=new StringBuffer();
        try { Class.forName("sun.jdbc.odbc.JdbcOdbcDriver"); }
        catch(ClassNotFoundException e){ }
    }
    public void setKeyWord(String s) {
        keyWord=s.trim();
        try{   byte bb[]=keyWord.getBytes("ISO-8859-1");
               keyWord=new String(bb,"gb2312");
        }
        catch(Exception e) { }
    }
    public String getKeyWord(){
        return keyWord;
    }
    public void setZiduan(String s) {
        ziduan=s.trim();
        try{   byte bb[]=ziduan.getBytes("ISO-8859-1");
               ziduan=new String(bb,"gb2312");
```

```
        }
        catch(Exception e){ }
    }
    public String getZiduan(){
        return ziduan;
    }
    public StringBuffer getQueryResult(){
        String condition="SELECT * FROM score Where "+ziduan+" Like '%"+keyWord+"%'";
        StringBuffer str=f(condition);
        return str;
    }
    public StringBuffer f(String condition) {
        StringBuffer str=new StringBuffer();
        Connection con;
        Statement sql;
        ResultSet rs;
        try {   con=DriverManager.getConnection("jdbc:odbc:mymoon","sa","sa");
            sql=con.createStatement();
            rs=sql.executeQuery(condition);
            str.append("<table border=1>");
            str.append("<th width=80>"+"学号");
            str.append("<th width=80>"+"姓名");
            str.append("<th width=80>"+"数学成绩");
            str.append("<th width=80>"+"物理成绩");
            str.append("<th width=80>"+"英语成绩");
            while(rs.next()){
                str.append("<tr>");
                str.append("<td>"+rs.getString(1)+"</td>");
                str.append("<td>"+rs.getString(2)+"</td>");
                str.append("<td>"+rs.getString(3)+"</td>");
                str.append("<td>"+rs.getString(4)+"</td>");
                str.append("<td>"+rs.getString(5)+"</td>");
                str.append("</tr>");
            }
            str.append("<table border=1>");
            con.close();
        }
        catch(SQLException e) { str.append(e); }
        return str;
    }
}
```

JSP 页面文件如下：

keyWord.jsp（效果如图 7.29 所示） 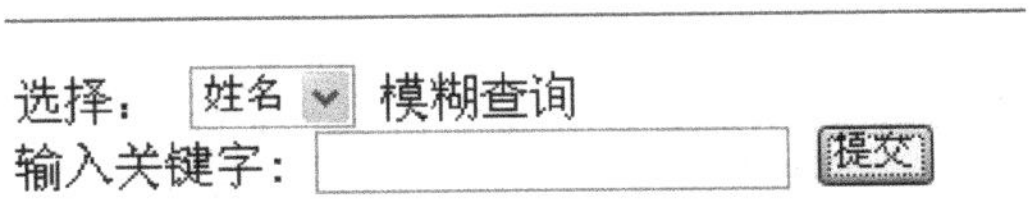

图 7.29 选择字段、输入关键字

```
<%@ page contentType="text/html;charset=GB2312" %>
<HTML><BODY><FONT size=3>
<FORM action="show.jsp" method=post name=form>
   选择:
    <Select name="ziduan">
       <Option value="姓名">姓名
       <Option value="学号">学号
    </Select>
   模糊查询
   <BR>输入关键字:
      <Input type="text" name="keyWord">
      <Input type=submit name="g" value="提交">
</FORM>
</BODY></HTML>
```

show.jsp (效果如图 7.30 所示)

"姓名"含有关键字 "小"的记录:

学号	姓名	数学成绩	物理成绩	英语成绩
0004	赵小林	89.0	79.0	89.0
0005	王近小	88.0	77.0	66.0

图 7.30 显示根据关键字查询的记录

```
<%@ page contentType="text/html;charset=GB2312" %>
<%@ page import="tom.jiafei.ConditionQuery" %>
<jsp:useBean id="database" class="tom.jiafei.KeyWordQuery" scope="request" />
<jsp:setProperty name="database"  property="ziduan" param="ziduan"/>
<jsp:setProperty name="database"  property="keyWord" param="keyWord"/>
<HTML><BODY>
   "<jsp:getProperty  name="database"  property="ziduan"  />"含有关键字
   "<jsp:getProperty name="database"  property="keyWord"  />"的记录:
<BR> <jsp:getProperty  name="database"  property="queryResult" />
</BODY></HTML>
```

7.5 更新、添加与删除操作

Statement 对象调用方法如下:

   public int executeUpdate(String sqlStatement);

通过参数 sqlStatement 指定的方式实现对数据库表中记录的更新、添加和删除操作。更新、添加和删除记录的 SQL 语法分别如下:

   UPDATE <表名> SET <字段名> = 新值 WHERE <条件子句>
   INSERT INTO score(字段列表) VALUES (对应的具体的记录)
或 INSERT INTO score(VALUES (对应的具体的记录)
   DELETE FROM <表名> WHERE <条件子句>

例如,下述 SQL 语句将 score 表中"姓名"字段值为'张三'的记录的"数学成绩"字段和物理成绩字段的值更新为 88 和 99:

UPDATE score SET 数学成绩 = 88,物理成绩=99 WHERE 姓名='张三'

下述 SQL 语句将向 score 表中添加一条新的记录（'0007','li',68,90,78）：

INSERT INTO score(学号,姓名,数学成绩,物理成绩,英语成绩) VALUES ('0007','li',68,90,78)

下述 SQL 语句将删除 score 表中的"学号"字段值为 0002 的记录：

DELETE FROM score WHERE 学号= '0002'

注意： 可以使用一个 Statement 对象进行更新操作，但当查询语句返回结果集后，没有立即输出结果集的记录，而接着执行了更新语句，那么结果集就不能输出记录了。要想输出记录就必须重新返回结果集。

【例 7-8】 通过 JSP 页面 mainPage.jsp 可以看到数据库 Student 的 score 表中的全部记录，在该页面单击"更新操作"、"添加操作"或"删除操作"可链接到相应的 JSP 页面 renew.jsp、add.jsp 和 delete.jsp，这些页面使用相应的 bean 完成"更新"、"添加"和"删除"操作。在这个例子中，除了新写的 bean 外，还用到了例 7-3 中的 bean(QueryBeanOne.java)，该 bean 帮助查询 score 表中的记录。

创建 bean 的源文件如下：

RenewBean.java

```java
package tom.jiafei;
import java.sql.*;
public class RenewBean{
    String number="",                              //学号
           name="";                                //姓名
    float  mathScore,physicsScore,englishScore;
    String renewMessage="";
    public RenewBean(){
       try { Class.forName("sun.jdbc.odbc.JdbcOdbcDriver"); }
       catch(ClassNotFoundException e){ }
    }
    public void setNumber(String s) {
       number=s.trim();
       try{ byte bb[]=number.getBytes("ISO-8859-1");
            number=new String(bb,"gb2312");
          }
       catch(Exception e){ }
    }
    public void setName(String s) {
       name=s.trim();
       try{ byte bb[]=name.getBytes("ISO-8859-1");
            name=new String(bb,"gb2312");
          }
       catch(Exception e){ }
    }
    public void setMathScore(float n) {
       mathScore=n;
    }
```

```java
            public void setPhysicsScore(float n) {
                physicsScore=n;
            }
            public void setEnglishScore(float n) {
                englishScore=n;
            }
            public String getRenewMessage(){
                String updateCondition="UPDATE score SET 姓名= '"+name+"'"+
                                    ", 数学成绩 = "+mathScore+", 物理成绩 = "+physicsScore+
                                    ", 英语成绩 = "+englishScore+"WHERE 学号 = '"+number+"'";
                String str="";
                Connection con;
                Statement sql;
                try {  con=DriverManager.getConnection("jdbc:odbc:mymoon","sa","sa");
                    sql=con.createStatement();
                    int m=sql.executeUpdate(updateCondition);
                    if(m!=0) {
                        str="对表中第"+m+"条记录更新成功";
                    }
                    else{
                        str="更新失败";
                    }
                    con.close();
                }
                catch(SQLException e) { str="你还没有提交更新的数据或"+e;  }
                return str;
            }
        }
```

AddBean.java

```java
        package tom.jiafei;
        import java.sql.*;
        public class AddBean {
            String number="",                   //学号
                   name="";                     //姓名
            float  mathScore,physicsScore,englishScore;
            String addMessage="";
            public AddBean(){
                try {  Class.forName("sun.jdbc.odbc.JdbcOdbcDriver");   }
                catch(ClassNotFoundException e) { }
            }
            public void setNumber(String s) {
                number=s.trim();
                try{   byte bb[]=number.getBytes("ISO-8859-1");
                    number=new String(bb,"gb2312");
                }
                catch(Exception e) { }
            }
```

```java
        public void setName(String s) {
            name=s.trim();
            try{   byte bb[]=name.getBytes("ISO-8859-1");
                    name=new String(bb,"gb2312");
            }
            catch(Exception e){ }
        }
        public void setMathScore(float n) {
            mathScore=n;
        }
        public void setPhysicsScore(float n) {
            physicsScore=n;
        }
        public void setEnglishScore(float n) {
            englishScore=n;
        }
        public String getAddMessage(){
            String insertCondition="INSERT score VALUES( '"+number+"', '"+name+"',"
                                    +mathScore+","+physicsScore+","+englishScore+")";
            String str="";
            Connection con;
            Statement sql;
            try {   con=DriverManager.getConnection("jdbc:odbc:mymoon","sa","sa");
                    sql=con.createStatement();
                    if(number.length()>0) {
                        int m=sql.executeUpdate(insertCondition);
                        if(m!=0) {
                            str="对表中添加"+m+"条记录成功";
                        }
                        else{
                            str="添加记录失败";
                        }
                    }
                    else{
                        str="必须要有学号";
                    }
                    con.close();
            }
            catch(SQLException e) {   str="输入的序号不允许重复"+e; }
            return str;
        }
    }
```

DelBean.java

```java
    package tom.jiafei;
    import java.sql.*;
    public class DelBean {
        String number="",                       //学号
```

```
            name="";                           //姓名
    float   mathScore,physicsScore,englishScore;
    String delMessage="";
    public DelBean(){
       try { Class.forName("sun.jdbc.odbc.JdbcOdbcDriver"); }
       catch(ClassNotFoundException e) { }
    }
    public void setNumber(String s) {
       number=s.trim();
       try{   byte bb[]=number.getBytes("ISO-8859-1");
              number=new String(bb,"gb2312");
       }
       catch(Exception e){ }
    }
    public String getDelMessage(){
       String delCondition="DELETE FROM score WHERE  学号 = "+"'"+number+"'";
       String str="";
       Connection con;
       Statement sql;
       try {  con=DriverManager.getConnection("jdbc:odbc:mymoon","sa","sa");
              sql=con.createStatement();
              if(number.length()>0) {
                  int m=sql.executeUpdate(delCondition);
                  if(m!=0) {
                     str="对表中删除"+m+"条记录成功";
                  }
                  else{
                     str="删除记录失败";
                  }
              }
              else{
                  str="必须指定要删除记录的学号";
              }
              con.close();
       }
       catch(SQLException e) { str="学号不存在"; }
       return str;
    }
}
```

JSP 页面文件如下：

mainPage.jsp（效果如图 7.31 所示）

```
<%@ page contentType="text/html;charset=GB2312" %>
<%@ page import="tom.jiafei.QueryBeanOne" %>
<jsp:useBean id="look" class="tom.jiafei.QueryBeanOne" scope="request" />
<jsp:setProperty   name= "look"    property="ODBCDataSource" value="mymoon" />
<jsp:setProperty   name= "look"    property="tableName" value="score" />
```

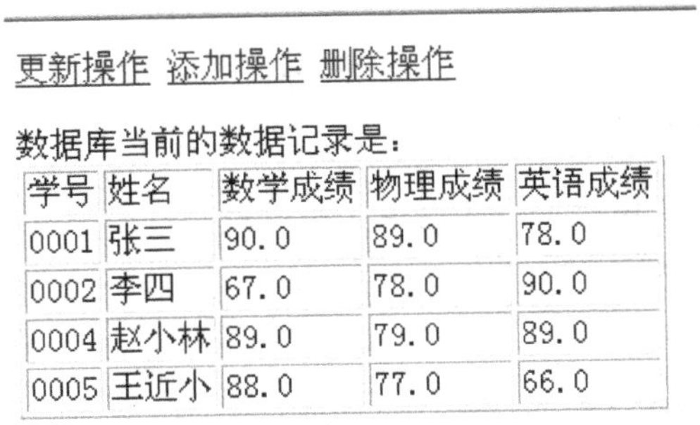

更新操作 添加操作 删除操作

数据库当前的数据记录是：

学号	姓名	数学成绩	物理成绩	英语成绩
0001	张三	90.0	89.0	78.0
0002	李四	67.0	78.0	90.0
0004	赵小林	89.0	79.0	89.0
0005	王近小	88.0	77.0	66.0

图 7.31 选择操作方式

```
<jsp:setProperty   name= "look"   property="user" value="sa" />
<jsp:setProperty   name= "look"   property="secret" value="sa" />
<HTML><BODY ><FONT size=3>
 <A href="renew.jsp">更新操作</A>
 <A href="add.jsp">添加操作</A>
 <A href="delete.jsp">删除操作</A>
<P>数据库当前的数据记录是：
 <jsp:getProperty   name= "look"   property="queryResult"   />
</BODY></HTML>
```

renew.jsp（效果如图 7.32 所示）

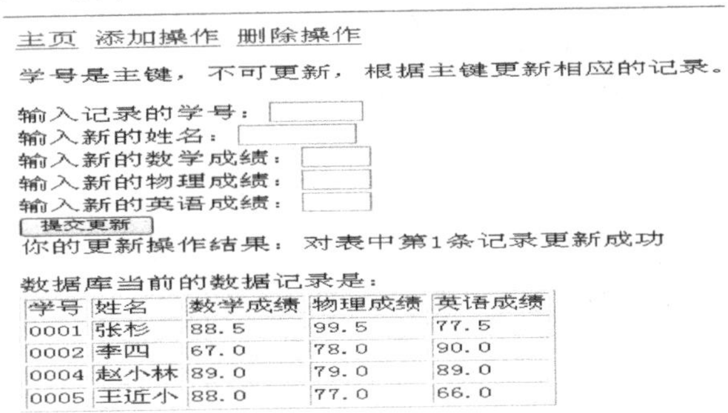

主页 添加操作 删除操作

学号是主键，不可更新，根据主键更新相应的记录。

输入记录的学号：
输入新的姓名：
输入新的数学成绩：
输入新的物理成绩：
输入新的英语成绩：
[提交更新]
你的更新操作结果：对表中第1条记录更新成功

数据库当前的数据记录是：

学号	姓名	数学成绩	物理成绩	英语成绩
0001	张杉	88.5	99.5	77.5
0002	李四	67.0	78.0	90.0
0004	赵小林	89.0	79.0	89.0
0005	王近小	88.0	77.0	66.0

图 7.32 更新操作

```
<%@ page contentType="text/html;charset=GB2312" %>
<%@ page import="java.sql.*" %>
<%@ page import="tom.jiafei.QueryBeanOne" %>
<%@ page import="tom.jiafei.RenewBean" %>
<jsp:useBean id="look" class="tom.jiafei.QueryBeanOne" scope="request" />
<jsp:useBean id="renew" class="tom.jiafei.RenewBean" scope="request" />
<jsp:setProperty   name= "look"   property="ODBCDataSource" value="mymoon" />
<jsp:setProperty   name= "look"   property="tableName" value="score" />
<jsp:setProperty   name= "look"   property="user" value="sa" />
<jsp:setProperty   name= "look"   property="secret" value="sa" />
<HTML><BODY ><FONT size=4>
 <A href="mainPage.jsp">主页</A>
 <A href="add.jsp">添加操作</A>
 <A href="delete.jsp">删除操作</A>
<P>学号是主键，不可更新，根据主键更新相应的记录。
```

```
<FORM action="" method=post>
 输入记录的学号：<Input type="text" name="number" size=6>
<BR>输入新的姓名：<Input type="text" name="name" size=8>
<BR>输入新的数学成绩：<Input type="text" name="mathScore" size=4>
<BR>输入新的物理成绩：<Input type="text" name="physicsScore" size=4>
<BR>输入新的英语成绩：<Input type="text" name="englishScore" size=4>
<BR><Input type="submit" name="b" value="提交更新">
<jsp:setProperty   name= "renew"   property="*" />
<BR>你的更新操作结果：
<jsp:getProperty   name= "renew"   property="renewMessage" />
<P>数据库当前的数据记录是：
 <jsp:getProperty   name= "look"   property="queryResult"   />
</FONT>
</BODY></HTML>
```

add.jsp（效果如图 7.33 所示）

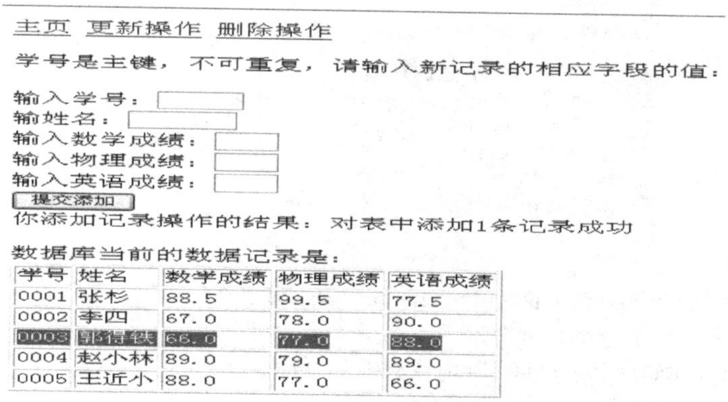

图 7.33 添加操作

```
<%@ page contentType="text/html;charset=GB2312" %>
<%@ page import="java.sql.*" %>
<%@ page import="tom.jiafei.QueryBeanOne" %>
<%@ page import="tom.jiafei.AddBean" %>
<jsp:useBean id="look" class="tom.jiafei.QueryBeanOne" scope="request" />
<jsp:useBean id="add" class="tom.jiafei.AddBean" scope="request" />
<jsp:setProperty   name= "look"   property="ODBCDataSource" value="mymoon" />
<jsp:setProperty   name= "look"   property="tableName" value="score" />
<jsp:setProperty   name= "look"   property="user" value="sa" />
<jsp:setProperty   name= "look"   property="secret" value="sa" />
<HTML><BODY ><FONT size=4>
  <A href="mainPage.jsp">主页</A>
  <A href="renew.jsp">更新操作</A>
  <A href="delete.jsp">删除操作</A>
<P>学号是主键，不可重复，请输入新记录的相应字段的值：
<FORM action="" method=post>
 输入学号：<Input type="text" name="number" size=6>
<BR>输姓名：<Input type="text" name="name" size=8>
<BR>输入数学成绩：<Input type="text" name="mathScore" size=4>
```

```
    <BR>输入物理成绩：<Input type="text" name="physicsScore" size=4>
    <BR>输入英语成绩：<Input type="text" name="englishScore" size=4>
    <BR><Input type="submit" name="b" value="提交添加">
    <jsp:setProperty  name="add"  property="*" />
    <BR>你添加记录操作的结果：
    <jsp:getProperty  name="add"  property="addMessage" />
    <P>数据库当前的数据记录是：
    <jsp:getProperty  name="look"  property="queryResult"  />
    </FONT>
    </BODY></HTML>
```

delete.jsp（效果如图 7.34 所示）

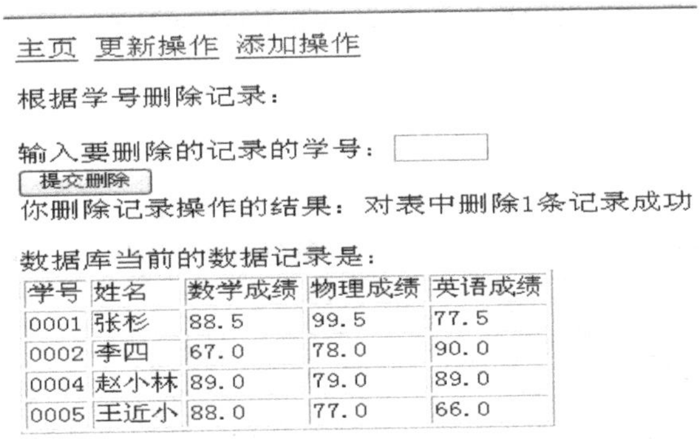

图 7.34 删除操作

```
<%@ page contentType="text/html;charset=GB2312" %>
<%@ page import="java.sql.*" %>
<%@ page import="tom.jiafei.QueryBeanOne" %>
<%@ page import="tom.jiafei.DelBean" %>
<jsp:useBean id="look" class="tom.jiafei.QueryBeanOne" scope="request" />
<jsp:useBean id="del" class="tom.jiafei.DelBean" scope="request" />
<jsp:setProperty  name="look"  property="ODBCDataSource" value="mymoon" />
<jsp:setProperty  name="look"  property="tableName" value="score" />
<jsp:setProperty  name="look"  property="user" value="sa" />
<jsp:setProperty  name="look"  property="secret" value="sa" />
<HTML><BODY><FONT size=4>
   <A href="mainPage.jsp">主页</A>
   <A href="renew.jsp">更新操作</A>
   <A href="add.jsp">添加操作</A>
   <P>根据学号删除记录：
   <FORM action="" method=post>
   输入要删除的记录的学号：<Input type="text" name="number" size=6>
   <BR><Input type="submit" name="b" value="提交删除">
   <jsp:setProperty  name="del"  property="*" />
   <BR>你删除记录操作的结果：
   <jsp:getProperty  name="del"  property="delMessage" />
   <P>数据库当前的数据记录是：
   <jsp:getProperty  name="look"  property="queryResult"  />
   </FONT></BODY></HTML>
```

7.6 使用预处理语句

Java 提供了更高效率的数据库操作机制，就是 PreparedStatement 对象，该对象被习惯地称为预处理语句对象。本节学习怎样使用预处理语句对象操作数据库中的表。

7.6.1 预处理语句的优点

当向数据库发送一个 SQL 语句，如 "SELECT * FROM score"，数据库中的 SQL 解释器负责将把 SQL 语句生成地层的内部命令，然后执行该命令，完成有关的数据操作。如果不断地向数据库提交 SQL 语句势必增加数据库中 SQL 解释器的负担，影响执行的速度。如果应用程序能针对连接的数据库，事先就将 SQL 语句解释为数据库地层的内部命令，然后直接让数据库去执行这个命令，显然不仅减轻了数据库的负担，而且也提高了访问数据库的速度。

对于 JDBC，若使用 Connection 和某个数据库建立了连接对象 con，那么 con 就可以调用

prepareStatement(String sql)

方法对参数 sql 指定的 SQL 语句进行预编译处理，生成该数据库地层的内部命令，并将该命令封装在 PreparedStatement 对象中，那么该对象调用下列方法都可以使得该地层内部命令被数据库执行：

ResultSet executeQuery()
boolean execute()
int executeUpdate()

只要编译好了 PreparedStatement 对象，那么该对象可以随时地执行上述方法，显然提高了访问数据库的速度。

【例 7-9】 bean 使用预处理语句来查询数据库表中的全部记录。

创建 bean 的源文件如下：

PreparedQueryBean.java
```
    package tom.jiafei;
    import java.sql.*;
    public class PreparedQueryBean {
        String databaseName="";              //数据库名称
        String tableName="";                 //表的名字
        String user=""          ;            //用户
        String password="" ;                 //密码
        StringBuffer queryResult;            //查询结果
        public PreparedQueryBean(){
            queryResult=new StringBuffer();
            try{ Class.forName("com.microsoft.sqlserver.jdbc.SQLServerDriver");   }
            catch(Exception e) {
                queryResult=new StringBuffer();
                queryResult.append(""+e);
            }
        }
        public void setDatabaseName(String s) {
            databaseName=s.trim();
```

```java
        }
        public String getDatabaseName(){
            return databaseName;
        }
        public void setTableName(String s) {
            tableName=s.trim();
        }
        public String getTableName(){
            return tableName;
        }
        public void setPassword(String s) {
            password=s.trim();;
        }
        public String getPassword(){
            return password;
        }
        public void setUser(String s) {
            user=s.trim();;
        }
        public String getUser(){
            return user;
        }
        public StringBuffer getQueryResult(){
            Connection con;
            PreparedStatement sql;
            ResultSet rs;
            try {   queryResult.append("<table border=1>");
                String uri= "jdbc:sqlserver://127.0.0.1:1433;DatabaseName="+databaseName;
                con=DriverManager.getConnection(uri,user,password);
                DatabaseMetaData metadata=con.getMetaData();
                ResultSet rs1=metadata.getColumns(null,null,tableName,null);
                int 字段个数=0;
                queryResult.append("<tr>");
                while(rs1.next()){
                    字段个数++;
                    String clumnName=rs1.getString(4);
                    queryResult.append("<td>"+clumnName+"</td>");
                }
                queryResult.append("</tr>");
                sql=con.prepareStatement("SELECT * FROM "+tableName); //预处理语句
                rs=sql.executeQuery();
                while(rs.next()){
                    queryResult.append("<tr>");
                    for(int k=1;k<=字段个数;k++) {
                        queryResult.append("<td>"+rs.getString(k)+"</td>");
                    }
                    queryResult.append("</tr>");
```

```
                }
                queryResult.append("</table>");
                con.close();
            }
            catch(SQLException e) {  queryResult.append(e); }
            return queryResult;
        }
    }
```

JSP 页面文件如下：

pre.jsp
```
<%@ page contentType="text/html;charset=GB2312" %>
<%@ page import="tom.jiafei.PreparedQueryBean" %>
<jsp:useBean id="database" class="tom.jiafei.PreparedQueryBean" scope="request" />
<HTML><BODY><FONT size=2>
<H5>使用预处理语句查询数据库：</H5>
<FORM action="" Method="post" >
    输入数据库的名字：<Input type=text name="databaseName" value="Student" size=8>
    输入表的名字： <Input type=text name="tableName" value="score" size=8>
    <BR>输入用户名称：<Input type=text name="user" size=6>
    输入密码：<Input type="password" name="password" size=6>
    <Input type=submit name="g" value="提交">
<jsp:setProperty  name= "database"   property="*" />
<BR> 在<jsp:getProperty  name= "database"  property="tableName"  />表查询到记录：
<BR> <jsp:getProperty  name= "database"  property="queryResult"  />
</FORM>
</BODY></HTML>
```

7.6.2 使用通配符

在对 SQL 进行预处理时，可以使用通配符 " ? " 来代替字段的值，只要在预处理语句执行之前再设置通配符所表示的具体值即可。例如：

  sql=con.prepareStatement("SELECT * FROM score WHERE 数学成绩 < ? ");

那么在 sql 对象执行之前，必须调用相应的方法设置通配符 " ? " 代表的具体值，例如：

  sql.setInt(1,89);

指定上述预处理 SQL 语句中通配符 " ? " 代表的值是 89。通配符按照预处理 SQL 语句中从左到右依次出现的顺序分别被称为第 1 个、第 2 个、…通配符，如下列方法：

  void setInt(int parameterIndex, int x)

用来设置通配符的值。参数 parameterIndex 用来表示 SQL 语句中从左到右的第 parameterIndex 个通配符号，x 是该通配符所代表的具体值。

  尽管

    sql=con.prepareStatement("SELECT * FROM score WHERE 数学成绩 < ? ");
    sql.setInt(1,89);

的功能等同于

    sql=con.prepareStatement("SELECT * FROM score WHERE 数学成绩 < 89 ");

但是，使用通配符可以使得应用程序更容易动态地改变 SQL 语句中关于字段值的条件。

预处理语句设置通配符"？"的值的常用方法有：

  void setDate(int parameterIndex,Date x)
  void setDouble(int parameterIndex,double x)
  void setFloat(int parameterIndex,float x)
  void setInt(int parameterIndex,int x)
  void setLong(int parameterIndex,long x)
  void setString(int parameterIndex,String x)

【例 7-10】 bean 使用预处理语句向 Student 数据库中的 score 表添加记录。本例还用到了 7.4.3 节中的 bean(ConditionQuery.java)。

创建 bean 的源文件如下：

AddRecordBean.java

```java
package tom.jiafei;
import java.sql.*;
public class AddRecordBean{
    String number="",                              //学号
           name="";                                //姓名
    float  mathScore,physicsScore,englishScore;
    String addMessage="";
    boolean ok=false;
    public AddRecordBean(){
        try { Class.forName("com.microsoft.sqlserver.jdbc.SQLServerDriver"); }
        catch(ClassNotFoundException e) { }
    }
    public void setNumber(String s) {
        number=s.trim();
        try{   byte bb[]=number.getBytes("ISO-8859-1");
               number=new String(bb,"gb2312");
        }
        catch(Exception e){ }
    }
    public void setName(String s) {
        name=s.trim();
        try{   byte bb[]=name.getBytes("ISO-8859-1");
               name=new String(bb,"gb2312");
        }
        catch(Exception e){ }
    }
    public void setMathScore(float n) {
        mathScore=n;
    }
    public void setPhysicsScore(float n) {
        physicsScore=n;
    }
    public void setEnglishScore(float n) {
```

```
                englishScore=n;
            }
            public boolean getOk(){
                return ok;
            }
            public String getAddMessage(){
                String str="";
                Connection con;
                PreparedStatement sql;
                try {   String uri= "jdbc:sqlserver://127.0.0.1:1433;DatabaseName=Student";
                    con=DriverManager.getConnection(uri,"sa","sa");
                    String insertCondition="INSERT INTO score VALUES (?,?,?,?,?)";
                    sql=con.prepareStatement(insertCondition);
                    if(number.length()>0) {
                        sql.setString(1,number);
                        sql.setString(2,name);
                        sql.setFloat(3,mathScore);
                        sql.setFloat(4,physicsScore);
                        sql.setFloat(5,englishScore);
                        int m=sql.executeUpdate();
                        if(m!=0) {
                            str="对表中添加"+m+"条记录成功";
                            ok=true;
                        }
                        else{
                            str="添加记录失败";
                        }
                    }
                    else{
                        str="必须要有学号";
                    }
                    con.close();
                }
                catch(SQLException e) { str="你还没有提供添加的数据或"+e; }
                return str;
            }
        }
```

JSP 页面文件如下：

pre.jsp

```
<%@ page contentType="text/html;charset=GB2312" %>
<%@ page import="java.sql.*" %>
<%@ page import="tom.jiafei.ConditionQuery" %>
<%@ page import="tom.jiafei.AddRecordBean" %>
<jsp:useBean id="look" class="tom.jiafei.ConditionQuery" scope="request" />
<jsp:useBean id="add" class="tom.jiafei.AddRecordBean" scope="request" />
<HTML><BODY ><FONT size=4>
    <P>学号是主键，不可重复，请输入新记录的相应字段的值：
```

```
<FORM action="" method=post>
   输入学号：<Input type="text" name="number" size=6>
<BR>输姓名：<Input type="text" name="name" size=8>
<BR>输入数学成绩：<Input type="text" name="mathScore" size=4>
<BR>输入物理成绩：<Input type="text" name="physicsScore" size=4>
<BR>输入英语成绩：<Input type="text" name="englishScore" size=4>
<BR><Input type="submit" name="b" value="提交添加">
<jsp:setProperty  name= "add"   property="*" />
<br>你添加记录操作的结果：
<jsp:getProperty  name= "add"   property="addMessage" />
<jsp:setProperty  name= "look"  property="number" param="number"/>
<P>添加的记录是：
 <%  if(add.getOk()==true)  {
 %>      <jsp:getProperty  name= "look"  property="queryResultByNumber" />
 <%  }
 %>
</FONT></BODY></HTML>
```

7.7 基于 CachedRowSet 分页显示记录

我们已经知道，Statement 对象将查询的结果返回到 ResultSet 对象中，但是 ResultSet 对象与数据库连接对象（Connnection 对象）实现了紧密的绑定，一旦连接对象被关闭，ResultSet 对象中的数据立刻消失。当用户使用分页方式显示表中记录时，就意味着必须始终保持与数据库的连接，直到用户将 ResultSet 对象中的数据查看完毕。我们知道，每种数据库在同一时刻都有允许的最大连接数目，因此当多个用户同时分页读取数据库表的记录时，应当避免长时间占用数据库的连接资源。

com.sun.rowset 包提供了 CachedRowSetImpl 类，该类实现了 CachedRowSet 接口。CachedRowSetImpl 对象可以保存 ResultSet 对象中的数据，而且 CachedRowSetImpl 对象不依赖 Connnection 对象，这意味着一旦把 ResultSet 对象中的数据保存到 CachedRowSetImpl 对象后，就可以关闭和数据库的连接。CachedRowSetImpl 继承了 ResultSet 的所有方法，因此可以像操作 ResultSet 对象一样来操作 CachedRowSetImpl 对象。将 ResultSet 对象 resultset 中的数据保存到 CachedRowSetImpl 对象 rowSet 中的代码如下：

```
CachedRowSetImpl rowSet=new CachedRowSetImpl ();
rowSet.populate(resultset);
```

假设 CachedRowSetImpl 对象中有 m 行记录，准备每页显示 n 行，那么总页数的计算规则如下：

- 如果 m 除以 n 的余数大于 0，总页数等于 m 除以 n 的商加 1。
- 如果 m 除以 n 的余数等于 0，总页数等于 m 除以 n 的商。

即总页数=(m%n)==0?(m/n):(m/n+1)。

如果准备显示第 p 页的内容，应当把 CachedRowSetImp1 对象中的游标移动到第 $(p-1)\times n+1$ 行记录处。

【例 7-11】 bean 在实现分页显示记录时使用了 CachedRowSet 对象。

创建 bean 的源文件：

ShowRecordByPage.java

```java
package tom.jiafei;
import java.sql.*;
import com.sun.rowset.*;
public class ShowRecordByPage{
    int pageSize=10;                              //每页显示的记录数
    int pageAllCount=0;                           //分页后的总页数
    int showPage=1        ;                       //当前显示页
    StringBuffer presentPageResult;               //显示当前页内容
    CachedRowSetImpl rowSet;                      //用于存储 ResultSet 对象
    String databaseName="";                       //数据库名称
    String tableName="";                          //表的名字
    String user=""        ;                       //用户
    String password="" ;                          //密码
    String 字段[]=new String[100]    ;
    int 字段个数=0;
    public ShowRecordByPage(){
        presentPageResult=new StringBuffer();
        try{ Class.forName("com.microsoft.sqlserver.jdbc.SQLServerDriver").newInstance(); }
        catch(Exception e) { }
    }
    public void setPageSize(int size) {
        pageSize=size;
        字段个数=0;
        String uri="jdbc:sqlserver://127.0.0.1:1433;DatabaseName="+databaseName;
        try{   Connection con=DriverManager.getConnection(uri,user,password);
            DatabaseMetaData metadata=con.getMetaData();
            ResultSet rs1=metadata.getColumns(null,null,tableName,null);
            int k=0;
            while(rs1.next()){
                字段个数++;
                字段[k]=rs1.getString(4); //获取字段的名字
                k++;
            }
            Statement sql=con.createStatement(ResultSet.TYPE_SCROLL_SENSITIVE,
                                        ResultSet.CONCUR_READ_ONLY);
            ResultSet rs=sql.executeQuery("SELECT * FROM "+tableName);
            rowSet=new CachedRowSetImpl();            //创建行集对象
            rowSet.populate(rs);
            con.close();                              //关闭连接
            rowSet.last();
            int m=rowSet.getRow();                    //总行数
            int n=pageSize;
             pageAllCount=((m%n)==0)?(m/n):(m/n+1);
        }
        catch(Exception exp) { }
```

```java
        }
        public int getPageSize(){
            return pageSize;
        }
        public int getPageAllCount(){
            return pageAllCount;
        }
        public void setShowPage(int n) {
            showPage=n;
        }
        public int getShowPage(){
           return showPage;
        }
        public StringBuffer getPresentPageResult(){
            if(showPage>pageAllCount)
                showPage=1;
            if(showPage<=0)
                showPage=pageAllCount;
            presentPageResult=show(showPage);
            return presentPageResult;
        }
        public StringBuffer show(int page) {
            StringBuffer str=new StringBuffer();
            str.append("<table border=1>");
            str.append("<tr>");
            for(int i=0;i<字段个数;i++){
                str.append("<th>"+字段[i]+"</th>");
            }
            str.append("</tr>");
            try{   rowSet.absolute((page-1)*pageSize+1);
                  for(int i=1;i<=pageSize;i++){
                      str.append("<tr>");
                      for(int k=1;k<=字段个数;k++){
                          str.append("<td>"+rowSet.getString(k)+"</td>");
                      }
                      str.append("</tr>");
                      rowSet.next();
                  }
            }
            catch(SQLException exp) { }
            str.append("</table>");
            return str;
        }
        public void setDatabaseName(String s) {
            databaseName=s.trim();
        }
        public String getDatabaseName(){
```

```java
            return databaseName;
        }
        public void setTableName(String s) {
            tableName=s.trim();
        }
        public String getTableName(){
            return tableName;
        }
        public void setPassword(String s) {
            password=s.trim();;
        }
        public void setUser(String s) {
            user=s.trim();
        }
        public String getUser(){
            return user;
        }
    }
```

JSP 页面：

choiceDatabase.jsp（效果如图 7.35 所示）

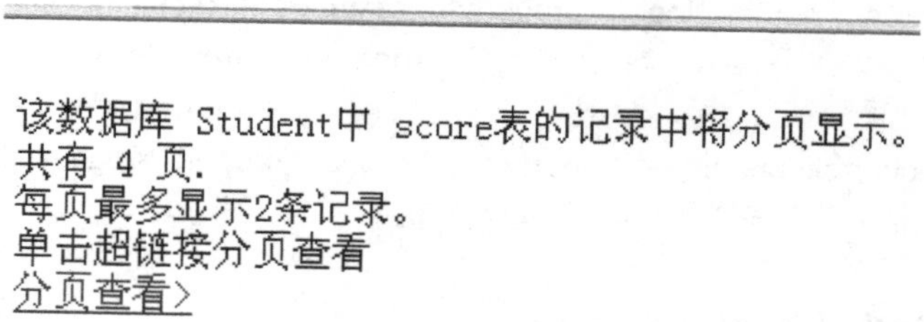

图 7.35　显示数据库分页结果

```jsp
<%@ page contentType="text/html;charset=GB2312" %>
<%@ page import="tom.jiafei.ShowRecordByPage" %>
<jsp:useBean id="look" class="tom.jiafei.ShowRecordByPage" scope="session" />
<jsp:setProperty name= "look"   property="databaseName" value="Student" />
<jsp:setProperty name= "look"   property="tableName" value="score" />
<jsp:setProperty name= "look"   property="user" value="sa" />
<jsp:setProperty name= "look"   property="password" value="sa" />
<jsp:setProperty name= "look"   property="pageSize" value="2" />
<BR> 该数据库
  <jsp:getProperty  name= "look"   property="databaseName"/>中
  <jsp:getProperty  name= "look"   property="tableName" />表
的记录中将分页显示。
<BR>共有 <jsp:getProperty  name= "look"  property="pageAllCount"  /> 页.
<BR>每页最多显示<jsp:getProperty  name= "look"  property="pageSize"  />条记录。
<BR>单击超链接分页查看
<BR><A href="showBypage.jsp">分页查看></A>
</BODY></HTML>
```

showBypage.jsp（效果如图 7.36 所示）

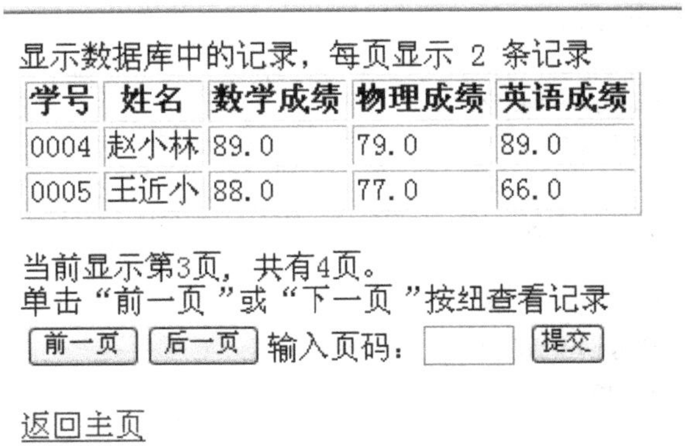

显示数据库中的记录，每页显示 2 条记录

学号	姓名	数学成绩	物理成绩	英语成绩
0004	赵小林	89.0	79.0	89.0
0005	王近小	88.0	77.0	66.0

当前显示第3页，共有4页。
单击"前一页"或"下一页"按纽查看记录
[前一页] [后一页] 输入页码：[   ] [提交]

返回主页

图 7.36 分页查看记录

```
<%@ page contentType="text/html;charset=GB2312" %>
<%@ page import="tom.jiafei.ShowRecordByPage" %>
<jsp:useBean id="look" class="tom.jiafei.ShowRecordByPage" scope="session" />
<P>显示数据库中的记录，每页显示
<jsp:getProperty   name= "look"   property="pageSize"   />
条记录
<jsp:setProperty   name= "look"   property="showPage"   />
<jsp:getProperty   name= "look"   property="presentPageResult" />
<BR>当前显示第<Font color=blue><jsp:getProperty   name= "look"
                                     property="showPage"   /></Font>页，
共有<Font color=blue><jsp:getProperty name= "look" property="pageAllCount" /></Font>页。
<BR>单击"前一页"或"下一页"按纽查看记录
<TABLE>
  <TR><TD><FORM action="">
           <Input type=hidden name="showPage" value="<%=look.getShowPage()-1 %>" >
           <Input type=submit name="g" value="前一页">
         </FORM>
      </TD>
      <TD><FORM action="">
           <Input type=hidden name="showPage" value="<%=look.getShowPage()+1 %>" >
           <Input type=submit name="g" value="后一页">
         </FORM>
      </TD>
      <TD> <FORM action="">
           输入页码：<Input type=text name="showPage" size=5 >
           <Input type=submit name="g" value="提交">
         </FORM>
      </TD>
  </TR>
</TABLE>
<A href="choiceDatabase.jsp">返回主页</A>
</BODY></HTML>
```

7.8 常见数据库的连接

7.8.1 连接 Oracle 数据库

和连接 SQL Server 数据库类似，我们也可以通过 JDBC-ODBC 桥接器和加载纯 Java 数据库驱动程序和 Oracle 数据库建立连接。JDBC-ODBC 桥接器方式的设置和 7.3.1 节介绍的类似，只是在为 ODBC 数据源选择驱动程序时，选择 Oracle 即可。下面介绍通过直接加载纯 Java 数据库驱动程序来连接 Oracle 数据库的方法。

安装 Oracle 后，找到目录 Oracle\ora8i\jdbc 中的文件 classes12.zip；将该文件复制到 JDK 安装目录的子目录\jre\lib\ext 中，并将 classes.zip 重新命名为 classes.jre 或 classes.jar。

通过如下两个步骤，可以与一个 Oracle 数据库建立连接。

<1> 加载驱动程序：
Class.forName("oracle.jdbc.driver.OracleDriver");

<2> 建立连接：
Connection conn=DriverManager.getConnection("jdbc:oracle:thin:
@主机 host:端口号:数据库名", "用户名", "密码");

【例 7-12】 JSP 页面连接 Oracle 数据库。

linkOracle.jsp

```
<%@ page contentType="text/html;charset=GB2312" %>
<%@ page import="java.sql.*" %>
<HTML>
<BODY>
<% Connection con=null;
   Statement sql=null;
   ResultSet rs=null;
   try{ Class.forName("oracle.jdbc.driver.OracleDriver");  }
   catch(ClassNotFoundException e){ }
   try { con=DriverManager.getConnection("jdbc:oracle:thin:
                       @192.168.0.35:1521:Lea", "scott","tiger");
      sql=con.createStatement();
      rs=sql.executeQuery("select * from emp");
      out.print("<Table Border>");
      out.print("<TR>");
      out.print("<TH width=100>"+"EMPNO");
      out.print("<TH width=50>"+"Ename");
      out.print("</TR>");
      while(rs.next()){
          out.print("<TR>");
          int n=rs.getInt(1);
          out.print("<TD >"+n+"</TD>");
          String e=rs.getString(2);
```

```
                    out.print("<TD >"+e+"</TD>");
                    out.print("</TR>") ;
            }
            out.print("</Table>");
            con.close();
        }
        catch(SQLException e1) { out.print(""+e1); }
%>
</BODY>
</HTML>
```

7.8.2 连接 MySql 数据库

目前，ODBC 不支持 MySql 数据库，因此不能使用 JDBC-ODBC 桥接器方式与 MySql 数据库建立连接，只能使用加载 MySql 的纯 Java 驱动程序来与 MySql 数据库建立连接。本节首先介绍 MySql 数据库服务器的安装与使用，然后介绍怎样与 MySql 数据库建立连接。

1．安装 MySql 数据库

MySql 是目前比较流行的一种网络数据库，是开源项目，很多网站都提供免费下载。可以使用任何搜索引擎搜索关键字："MySql 下载"，来获得有关的下载地址。我们选择的地址是 MySqL 的官方网站 http://www.mysql.com，该网站免费提供 MySQL 最新版本的下载以及相关技术文章。我们下载的版本是 mysql-5.0.22-win32.zip，该版本可以安装在 Windows 操作系统平台上（如果是其他操作系统平台，请按网站的提示下载相应的版本）。

mysql-5.0.22-win32.zip 的安装步骤如下：

<1> 将 mysql-5.0.22-win32.zip 解压到某个目录，如 E:\mysql。

<2> 双击 E:\mysql 下的 setup.exe 开始安装。出现安装界面，在该界面上单击【next】。

<3> 出现选择安装方式界面。有三种方式供选择：Typical、Complete 和 Custom，默认安装方式是 Typical。我们选择 Custom 方式，因为 Typical 和 Complete 方式不允许改变安装路径，只能安装在 C 盘上。

<4> 出现选择 MySql 功能以及安装路径界面。由于我们需要 MySQL 是一个网络数据库，即一个数据库服务器，所以选择"MySql Server"，这也是默认选择。单击该界面上的"Change"按钮，选择安装路径，我们选择的安装路径是 "D:\MySQLServer5.0"。

<5> 出现确认安装界面。开始安装过程，会出现安装进度条，只要按界面提示进行操作即可。

<6> 出现是否需要在 mysql.com 上注册的界面，我们选择跳过，即 "Skip Sign-Up"，然后单击 "Next" 按钮

<7> 出现确认安装结束界面。将该界面上是否配置 MySql 选择为否，即使用默认配置，单击 "确认" 按钮，完成安装过程。

2．启动 MySql 数据库服务器

MySql 数据库安装后会有如图 7.37 所示的目录结构。

为了启动 MySql 数据库服务器，需要执行 MySql 安装目录的 bin 子目录中的 mysqld.exe

文件。你需要打开 MS-DOS 命令行窗口，并使用 MS-DOS 命令 "cd D:\MySQL Server 5.0\bin" 进入到 bin 目录中，然后在命令行输入 "mysqld-nt" 启动 MySql 数据库服务器，如果启动成功，MySql 数据库服务器将占用当前 MS-DOS 窗口，如图 7.38 所示。

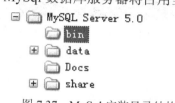

图 7.37 MySql 安装目录结构 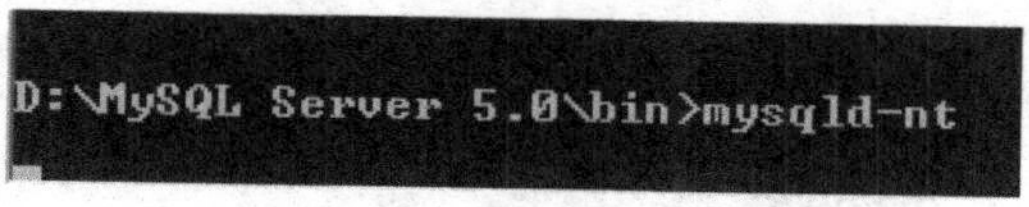　　　　　图 7.38 启动 MySql 服务器

3. 启动 MySql 监视程序

通过启动 MySql 监视程序，可以实现创建数据库、建表等操作。为了启动 MySql 监视程序，需要执行 MySql 安装目录的 bin 子目录中的 mysql.exe 文件，执行格式为：

　　mysql –u 用户名 密码

你需要再一个打开 MS-DOS 命令行窗口，并使用 MS-DOS 命令进入到 bin 目录中，然后使用默认的 root 用户登录（在安装时 root 用户是默认的一个用户，没有密码）。命令如下：

　　mysql –u root

成功登录后，MS-DOS 窗口出现 "mysql>" 字样的登录效果，如图 7.39 所示。如果想退出 MySql 监视程序，输入 "exit" 即可。

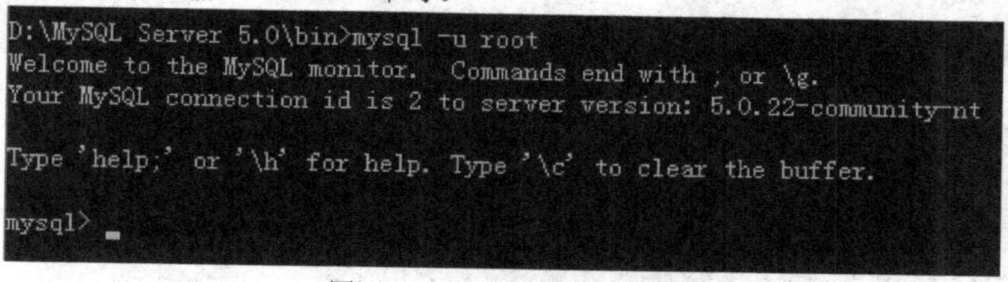

图 7.39 登录 MySql 服务器

4. 创建数据库

登录 MySql 服务器后，就可以使用 SQL 语句创建数据库、建表等操作。也可以下载相应的图形界面的 MySql 管理工具进行创建数据库、建表等操作，这些 MySql 管理工具有免费的，也有需要购买的。本节将讲述怎样在 JSP 中和 MySql 建立连接，所以在 MS-DOS 命令行窗口中输入 SQL 语句建立数据库和创建表。MySql 要求 SQL 语句必须用 "；" 号结束，在编辑 SQL 语句的过程中，可以使用 "\c" 终止当前 SQL 语句的编辑。

下面我们在登录后的 MS-DOS 窗口界面创建一个名字为 Car 的数据库，如图 7.40 所示。

5. 为数据库建表

创建数据库后就可以使用 SQL 语句在该库中进行建表操作。为了在某个数据库中建表，必须首先进入该数据库，命令为：

　　user 数据库名

进入数据库 Car 的操作如图 7.41 所示。

```
mysql> create database Car;
Query OK, 1 row affected (0.44 sec)

mysql>
```

图 7.40　创建数据库

```
mysql> use Car
Database changed
mysql>
```

图 7.41　进入数据库

下面建立一个名为 message 的表，该表的字段为：

number(varchar)
name(varchar)
madeTime(date)
price(float)

建表操作如图 7.42 所示。

建表之后，就可以使用 SQL 语句对表进行添加、更新和查询操作。现在我们向表 message 中添加 3 条记录，并使用 SQL 语句查询我们添加到表中的记录。操作过程如图 7.43 和图 7.44 所示。

```
mysql> create table message(
    -> number varchar(20) not null,
    -> name varchar(15) not null,
    -> madeTime date,
    -> price float,
    -> Primary Key (number)
    -> );
```

图 7.42　在数据库中建表

```
mysql> select * from message;
+--------+----------+------------+--------+
| number | name     | madeTime   | price  |
+--------+----------+------------+--------+
| AA108  | Beijing  | 2008-10-10 |  98988 |
| BB158  | Shanghai | 2007-10-10 |  78988 |
| CC158  | Dalian   | 2008-12-31 | 100000 |
+--------+----------+------------+--------+
```

图 7.44　从表中查询记录

```
mysql> insert into message values('AA108','Beijing','2008-10-10',98988);
Query OK, 1 row affected (0.47 sec)

mysql> insert into message values('BB158','Shanghai','2007-10-10',78987.98);
Query OK, 1 row affected (0.42 sec)

mysql> insert into message values('CC158','Dalian','2008-12-31',99999.98);
Query OK, 1 row affected (0.00 sec)
```

图 7.43　向表中添加记录

6．使用纯 Java 数据库驱动程序连接 MySql 数据库

为了能和 MySql 数据库服务器管理的数据库建立连接，必须保证该 MySql 数据库服务器已经启动，如果你没有更改过 MySql 数据库服务器的配置，那么该数据库服务器占用的端口是 3306。

使用纯 Java 数据库驱动程序方式与数据库建立连接需要经过加载纯 Java 驱动程序，然后再和指定的数据库建立连接 2 个步骤。

<1> 加载纯 Java 驱动程序。

可以登录 MySql 的官方网站 http:www.mysql.com，下载驱动程序。我们下载的是：mysql-connector-java-5.0.4.zip，将该 ZIP 文件解压至硬盘，在解压目录下的 mysql-connector-java-5.0.4-bin.jar 文件就是连接 MySql 数据库的纯 Java 驱动程序。将该驱动程序复制到 Tomcat 服务器所使用的 JDK 的 \jre\lib\ext 文件夹中，如 D:\jdk1.5\jre\lib\ext，或复制到

Tomcat 服务器安装目录的\common\lib 文件夹中，如 D:\apache-tomcat-5.5.20\common\lib。

应用程序加载 MySql 驱动程序代码如下：

```
try{
        Class.forName("com.mysql.jdbc.Driver")
}
catch(Exception e){ }
```

<2> 与指定的数据库建立连接。

假设应用程序和 MySql 服务器在同一台计算机上，那么应用程序和数据库 Car 建立连接的代码如下：

```
try{    String uri= "jdbc:mysql://localhost/Car";
        String user="root";
        String password="123";
        con=DriverManager.getConnection(uri,user,password);
}
catch(SQLException e) { }
```

其中，root 用户有权访问数据库 Car，root 用户的密码是 123。如果 root 用户没有设置密码，那么将上述

```
String password="123";
```

更改为：

```
String password="";
```

【例 7-13】 JSP 页面连接 MySql 数据库 Car、查询 message 表。

linkMySql.jsp

```
<%@ page contentType="text/html;charset=GB2312" %>
<%@ page import="java.sql.*" %>
<HTML><BODY>
 <%  Connection con;
     Statement sql;
     ResultSet rs;
     try { Class.forName("com.mysql.jdbc.Driver");  }
     catch(Exception e) { out.print(e); }
     try {   String uri= "jdbc:mysql://localhost/Car";
             con=DriverManager.getConnection(uri,"root","");
             sql=con.createStatement();
             rs=sql.executeQuery("SELECT * FROM message ");
             out.print("<table border=2>");
             out.print("<tr>");
             out.print("<th width=100>"+"number");
             out.print("<th width=100>"+"name");
             out.print("<th width=50>"+"madeTime");
             out.print("<th width=50>"+"price");
             out.print("</TR>");
             while(rs.next()) {
                 out.print("<tr>");
```

```
                    out.print("<td >"+rs.getString(1)+"</td>");
                    out.print("<td >"+rs.getString(2)+"</td>");
                    out.print("<td >"+rs.getDate(3)+"</td>");
                    out.print("<td >"+rs.getFloat(4)+"</td>");
                    out.print("</tr>") ;
                }
                out.print("</table>");
                con.close();
            }
            catch(SQLException e1) { out.print(e1);   }
        %>
        </BODY></HTML>
```

【例 7-14】 bean 加载纯 Java 数据库驱动程序连接数据库、查询表。用户可以通过 JSP 页面 main.jsp 输入数据库名和表名（可以在 MySql 安装目录的子目录 data 中查看数据库以及数据库中的表，即扩展名为 .frm 的文件）。

创建 bean 的源文件如下：

AAA.java

```java
package tom.jiafei;
import java.sql.*;
public class AAA {
    String databaseName="";              //数据库名称
    String tableName="";                 //表的名字
    String user=""       ;               //用户
    String secret="" ;                   //密码
    StringBuffer queryResult;            //查询结果
    public AAA(){
        queryResult=new StringBuffer();
        try{ Class.forName("com.mysql.jdbc.Driver"); }
        catch(Exception e){ }
    }
    public void setDatabaseName(String s) {
        databaseName=s.trim();
    }
    public String getDatabaseName() {
        return databaseName;
    }
    public void setTableName(String s) {
        tableName=s.trim();
    }
    public String getTableName() {
        return tableName;
    }
    public void setSecret(String s) {
        secret=s.trim();
```

```
            }
            public void setUser(String s) {
                user=s.trim();
            }
            public StringBuffer getQueryResult() {
                Connection con;
                Statement sql;
                ResultSet rs;
                try {   queryResult.append("<table border=1>");
                        String uri= "jdbc:mysql://127.0.0.1/"+databaseName;
                        String id=user;
                        String password=secret;
                        con=DriverManager.getConnection(uri,id,password);
                        DatabaseMetaData metadata=con.getMetaData();
                        ResultSet rs1=metadata.getColumns(null,null,tableName,null);
                        int 字段个数=0;
                        queryResult.append("<tr>");
                        while(rs1.next()) {
                            字段个数++;
                            String clumnName=rs1.getString(4);
                            queryResult.append("<td>"+clumnName+"</td>");
                        }
                        queryResult.append("</tr>");
                        sql=con.createStatement();
                        rs=sql.executeQuery("SELECT * FROM "+tableName);
                        while(rs.next()){
                            queryResult.append("<tr>");
                            for(int k=1;k<=字段个数;k++) {
                                queryResult.append("<td>"+rs.getString(k)+"</td>");
                            }
                            queryResult.append("</tr>");
                        }
                        queryResult.append("</table>");
                        con.close();
                }
                catch(SQLException e) { queryResult.append(e); }
                return queryResult;
            }
        }
```

JSP 页面文件如下：

main.jsp（效果如图 7.45 所示）

```
<%@ page contentType="text/html;charset=GB2312" %>
<%@ page import="tom.jiafei.AAA" %>
<jsp:useBean id="database" class="tom.jiafei.AAA" scope="request" />
<HTML><BODY ><FONT size=2>
```

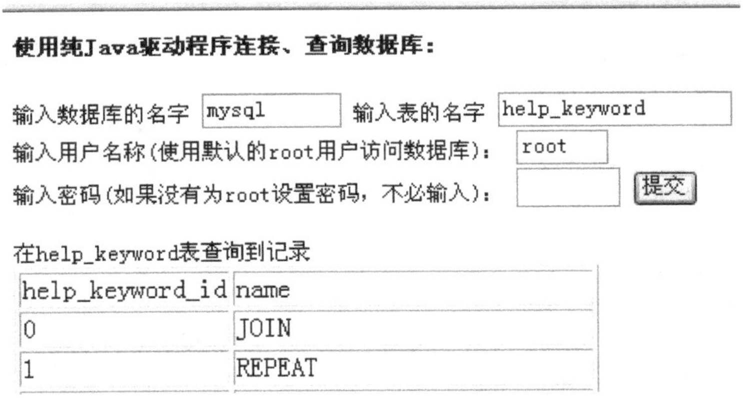

图 7.45 查询 MySql 数据库

```
<H5>使用纯 Java 驱动程序连接、查询数据库:</H5>
<FORM action=""  Method="post" >
   输入数据库的名字
   <Input type=text name="databaseName"   size=10>
   输入表的名字
   <Input type=text name="tableName" size=18>
   <BR>输入用户名称(使用默认的 root 用户访问数据库):
   <Input type=text name="user" size=6>
   <BR> 输入密码(如果没有为 root 设置密码,不必输入):
   <Input type="password" name="secret" size=6>
   <Input type=submit name="g" value="提交">
</FORM>
<jsp:setProperty   name= "database"   property="*" />
   在<jsp:getProperty   name= "database"   property="tableName"   />表查询到记录
     <jsp:getProperty   name= "database"   property="queryResult"   />
</BODY></HTML>
```

习 题 7

1. ODBC 设置数据源的主要步骤有哪些？
2. 参照例 7-1，编写一个查询 Access 数据库的 JSP 页面。
3. 加载纯 Java SQL Server 2000 数据库驱动程序的代码是什么？
4. 加载纯 Java MySql 数据库驱动程序的代码是什么？
5. 加载纯 Java Oracle 数据库驱动程序的代码是什么？
6. 使用 CachedRowSetImpl 类有什么好处？

第 8 章

Java Servlet 基础

本章导读

- 知识点：理解 Servlet 的工作原理以及生命周期。掌握怎样编写和使用 Servlet。编写和配置和 Servlet 有关的 web.xml 文件。
- 重点：理解和掌握 init()、service()、doPost() 和 doGet() 方法的使用。
- 难点：在 Servlet 中使用 session，使用 RequestDispatcher 把用户对某 JSP 页面或 Servlet 的请求转发给另一个 JSP 页面或 Servlet。
- 关键实践：编写 JSP 页面，使用 Servlet 处理有关数据。

在第 2 章我们曾学习了 JSP 运行原理，其核心内容是，当用户请求服务器上的一个 JSP 页面时，服务器负责把 JSP 页面文件转译成一个 Java 文件，再将这个 Java 文件编译生成字节码文件，然后将这个字节码文件加载到内存，并通过执行它来响应客户的请求。那么，服务器是怎样执行字节码的呢？实际上，服务器将字节码文件加载到内存后，就用该字节码创建了一个对象，然后让该对象响应用户的请求，当多个用户请求同一个 JSP 页面时，服务器会启动多个线程，在每个线程中，该对象响应用户的请求。

Java Servlet 技术就是在服务器端创建响应用户请求对象的技术，被创建的对象习惯上称为一个 Servlet 对象。在 JSP 技术出现之前，Web 应用开发人员就是自己编写创建 Servlet 对象的类，负责编译生成字节码文件，并复制该字节码文件到服务器的特定目录中，以便服务器使用该字节码创建一个 Servlet 对象来响应用户的请求。

JSP 技术就是以 Java Servlet 为基础，提供了 Java Servlet 的几乎所有好处，当客户请求一个 JSP 页面时，Tomcat 服务器自动生成 Java 文件、编译 Java 文件，并用编译得到的字节码文件在服务器端创建一个 Servlet 对象。但是 JSP 技术不是 Servlet 技术的全部，它只是 Servlet 技术的一个成功应用。JSP 技术屏蔽了 Servlet 对象创建的过程，使得 Web 程序设计者只需关心 JSP 页面本身的结构、设计好各种标记，如使用 HTML 标记设计页面的视图，使用 JavaBean 标记有效地分离页面的视图和数据处理等。

有些 Web 应用可能只需要 JSP+JavaBean 就能设计得很好，但是有些 Web 应用就可能需要 JSP+JavaBean+Servlet 来完成，即需要服务器再创建一些 Servlet 对象，配合 JSP 页面来完成整个 Web 应用程序的工作。关于这一点，我们将在第 9 章 MVC 模式中讲述。

8.1 Servlet 对象的创建与使用

本节将介绍怎样编写创建 Servlet 对象的类，怎样在 Tomcat 服务器上保存编译后的字节码、编写部署文件，怎样请求 Tomcat 服务器创建一个 Servlet 对象。有关 Servlet 对象的使用细节将在后续章节中讲述。

本章将使用 javax.servlet.http 包中的类，javax.servlet.http 包不在 JDK 的核心类库中，因此需要将 Tomcat 安装目录 lib 子目录中的 servlet-api.jar 文件复制到 Tomcat 服务器所使用的 JDK 的扩展目录中，如复制到 D:\jdk1.6\jre\lib\ext 中。

8.1.1 HttpServlet 类

熟悉 Java 语言的人更喜欢 Java Servlet，因为编写一个创建 Servlet 对象的类就是编写一个特殊类的子类，这个特殊的类就是 javax.servlet.http 包中的 HttpServlet 类。HttpServlet 类实现了 Servlet 接口，实现了响应用户的方法，这些方法将在后续内容中讲述。HttpServlet 类的子类被习惯地称为一个 Servlet 类。以下是一个简单的 Servlet 类，该类创建的 Servlet 可以响应用户的请求，即用户请求这个 Servlet 时，会在浏览器看到"您好，欢迎您。Hello,You are Welcome"这样的响应信息。

Servlet 源文件如下：

Hello.java
```
package star.moon;
import java.io.*;
import javax.servlet.*;
import javax.servlet.http.*;
public class Hello extends HttpServlet{
    public void init(ServletConfig config) throws ServletException {
        super.init(config);
    }
    public void service(HttpServletRequest reqest,HttpServletResponse response)
                                                      throws IOException {
        response.setContentType("text/html;charset=GB2312");   //设置响应的 MIME 类型
        PrintWriter out=response.getWriter();//获得一个向客户发送数据的输出流
        out.println("<HTML><BODY>");
        out.println("<H2>您好，欢迎您。Hello,You are Welcome</H2>");
        out.println("</BODY></HTML>");
    }
}
```

8.1.2 部署 Servlet

1. 字节码的保存

首先将 Servlet 源文件编译生成字节码文件，如上述 Servlet 类编译后的字节码文件就是

Hello.class。为了能让 Tomcat 服务器使用 Hello.class 字节码创建一个 Servlet 对象，需要将 Hello.class 保存到某个 Web 服务目录中的子目录中。

本章使用的 Web 服务目录是 chapter8。chapter8 是在 Tomcat 安装目录的 webapps 目录下建立的 Web 服务目录。

现在需要在当前 Web 服务目录下建立目录结构 chapter8\WEB-INF\classes，然后根据类的包名，在 classes 下再建立相应的子目录，如 Servlet 类的包名为 star.moon，那么在 classes 下建立子目录：\star\moon。为了让 Tomcat 服务器启用上述目录，必须重新启动 Tomcat 服务器。

Tomcat 服务器重新启动后，就可以将 Servlet 类的字节码，如 Hello.class，复制到目录 chapter8\WEB-INF\classes\star\moon 中。

2．编写部署文件

为了能让 Tomcat 服务器用我们的字节码文件创建对象，我们必须为 Tomcat 编写一个部署文件。该部署文件是一个 XML 文件，名字是 web.xml，由 Tomcat 服务器负责管理。这里不需深刻理解 XML 文件，只需知道 XML 文件是由标记组成的文件，使用该 XML 文件的应用程序（如 Tomcat 服务器）配有内置的解析器，可以解析 XML 标记中的数据。可以在 Tomcat 服务器的 WebApps 目录中的 root 目录找到一个 web.xml 文件，参照它编写自己的 web.xml 文件。

编写的 web.xml 文件保存到 Web 服务目录的 WEB-INF 子目录中，如 chapter8\WEB-INF。我们编写的 web.xml 文件的内容如下（需要用纯文本编辑）：

web.xml

```
<?xml version="1.0" encoding="ISO-8859-1"?>
<web-app>
    <servlet>
        <servlet-name>hello</servlet-name>
        <servlet-class>star.moon.Hello</servlet-class>
    </servlet>
    <servlet-mapping>
        <servlet-name>hello</servlet-name>
        <url-pattern>/lookHello</url-pattern>
    </servlet-mapping>
</web-app>
```

一个 XML 文件应当以 XML 声明作为文件的第一行，在其前面不能有空白、其他处理指令或注释。XML 声明以"<?xml"标识开始、以"?>"标识结束。注意："<?"和"xml"之间，以及"？"和">"之间不要有空格。如果在 XML 声明中没有显示地指定 encoding 属性的值，那么该属性的默认值为 UTF-8 编码。如果在编写 XML 文件时只准备使用 ASCII 字符，也可以将 encoding 属性的值设置为"ISO-8859-1"。例如：

```
<?xml version="1.0" encoding=" ISO-8859-1" ?>
```

这时，XML 文件必须使用 ANSI 编码保存（如图 8.1 所示），Tomcat 服务器中的 XML 解析器根据 encoding 属性的值也会识别 XML 文件中的标记，并正确解析标记中的内容。

如果 encoding 属性的值为 UTF-8，XML 文件必须按照 UTF-8 编码来保存（如图 8.2 所示）。如果 XML 使用 UTF-8 编码，那么标记以及标记的内容就可以使用汉字、日文、英文等，Tomcat 服务器中的 XML 解析器就会识别这些标记并正确解析标记中的内容。

图 8.1 encoding 值是 ISO-8859-1 时 XML 文件的保存

图 8.2 encoding 值是 UTF-8 时 XML 文件的保存

现在来看看 web.xml 文件中标记的具体内容及其作用。

XML 文件必须有一个根标记，web.xml 文件的根标记是<web-app>。web.xml 文件中可以有若干个<servlet>标记，该标记的内容由 Tomcat 服务器负责处理。<servlet>标记需要有 2 个子标记：<servlet-name>和<servlet-class>。其中，<servlet-name>标记的内容是 Tomcat 服务器创建的 Servlet 对象的名字。

web.xml 文件可以有若干个<servlet>标记，但要求各<servlet-name>的子标记的内容互不相同；<servlet-class>标记的内容指定 Tomcat 服务器用哪个类来创建 Servlet 对象。web.xml 文件中出现一个<servlet>标记就会对应地出现一个<servlet-mapping>标记，<servlet-mapping>有两个子标记：<servet-name>和<url-pattern>。其中，<servet-name>标记的内容是 Tomcat 服务器创建的 Servlet 对象的名字（该名字必须与<servlet>标记的子标记<servlet-name>标记的内容相同）；<servlet-mapping>标记用来指定用户用怎样的模式来请求 Servlet 对象，如<servlet-mapping>标记的内容是"/lookHello"，那么用户必须通过当前 Web 服务目录来请求 Servlet，如用户要请求服务器运行 Servlet 对象 hello 为其服务，那么在浏览器的地址栏中输入"http://127.0.0.1:8080/chapter8/lookHello"。

一个 Web 服务目录的 web.xml 文件负责管理该目录下的 Servlet 对象，当该 web 服务目录需要提供更多的 Servlet 对象时，只要在 web.xml 文件中增加<servlet>和<servlet-mapping>标记即可。

如果修改并重新保存 web.xml 文件，Tomcat 服务器就会立刻重新读取 web.xml 文件。因此，修改 web.xml 文件不必重新启动 Tomcat 服务器。但是，如果你的修改导致 web.xml 文件出现错误，Tomcat 服务器就会关闭当前 Web 服务目录下的所有 Servlet 的使用权限，直到保证 web.xml 文件正确无误。

8.1.3 运行 Servlet

用户可以根据 web.xml 部署文件来请求服务器执行一个 Servlet 对象。如果服务器没有名字为 hello 的 Servlet 对象，服务器就会根据 web.xml 文件中的<servlet-class>标记，用指定的类创建一个名字为 hello 的 Servlet 对象。因此，当 Servlet 对象 hello 被创建之后，如果修改了创建 Servlet 的 Java 源文件，并希望服务器用新的 Java 文件生成的字节码重新创建 Servlet 对象 hello，就必须重新启动 Tomcat 服务器。

Servlet 类可以使用 getServletName()方法返回配置文件中<servlet-name>标记给出的 Servlet 的名字。

当用户请求服务器运行一个 Servlet 对象时，需根据 web.xml 文件中<servlet-mapping>标记指定的格式输入请求，例如，用户请求服务器运行 Servlet 对象 hello 的效果如图 8.3 所示。

您好，欢迎您。Hello,You are Welcome

图 8.3　请求 Servlet 对象

8.2　Servlet 工作原理

Servlet 由支持 Servlet 的服务器负责管理运行，如 Tomcat 服务器。当多个客户请求一个 Servlet 时，服务器为每个客户启动一个线程而不是启动一个进程，这些线程由服务器来管理，与传统的 CGI 为每个客户启动一个进程相比较，效率要高得多。

1．Servlet 的生命周期

我们已经知道，一个 Servlet 对象是 javax.servlet 包中 HttpServlet 类的子类的一个实例、由服务器负责创建并完成初始化工作。一个 Servlet 对象的生命周期主要由下列 3 个过程组成：

<1> 初始化 Servlet 对象。Servlet 对象第一次被请求加载时，服务器初始化这个 Servlet 对象，即创建一个 Servlet 对象，对象调用 init()方法完成必要的初始化工作。

<2> 诞生的 Servlet 对象再调用 service()方法响应客户的请求。

<3> 当服务器关闭时，调用 destroy()方法，消灭 Servlet 对象。

Init()方法只被调用一次，即在 Servlet 第一次被请求加载时调用该方法。当后续的客户请求 Servlet 服务时，Web 服务将启动一个新的线程，在该线程中，Servlet 对象调用 service()方法响应客户的请求。也就是说，每个客户的每次请求都导致 service()方法被调用执行，分别运行在不同的线程中。

2．init()方法

init()方法是 HttpServlet 类中的方法，我们可以在 Servlet 类中重写这个方法。init()方法的声明格式如下：

　　public void init(ServletConfig config) 　throws ServletException

Servlet 对象第一次被请求加载时，服务器初始化一个 Servlet，即创建一个 Servlet 对象，它调用 init()方法完成必要的初始化工作。init()方法在执行时，服务器会把一个 SevletConfig 类型的对象传递给它，该对象就被保存在 Servlet 对象中，直到 Servlet 对象被消灭，该对象负责向 Servlet 传递服务设置信息，如果传递失败，就会发生 ServeletException，Servlet 对象就不能正常工作。

我们已经知道，当多个客户请求一个 Servlet 时，引擎为每个客户启动一个线程，那么 Servlet 类的成员变量被所有的线程共享。

3．service()方法

service()方法是 HttpServlet 类中的方法，我们可以在 Servlet 类中直接继承该方法或重写该方法。service()方法的声明格式如下：

```
public void service(HttpServletRequest request HttpServletResponse response)
                                              throw ServletException,IOException
```

当 Servlet 对象成功创建和初始化之后，该对象就调用 service()方法来处理用户的请求并返回响应。服务器将两个参数传递给该方法：一个 HttpServletRequest 类型的对象，该对象封装了用户的请求信息；另一个参数是 HttpServletResponse 类型的对象，该对象用来响应用户的请求。init()方法只被调用一次，而 service()方法可能被多次调用。我们已经知道，当后续客户请求该 Servlet 对象服务时，服务器将启动一个新的线程，在该线程中，Servlet 对象调用 service()方法响应客户的请求。也就是说，每个客户的每次请求都导致 service()方法被调用执行，调用过程运行在不同的线程中，互不干扰。因此，不同线程的 service()方法中的局部变量互不干扰，一个线程改变了自己的 service()方法中局部变量的值不会影响其他线程的 service()方法中的局部变量。

4．destroy()方法

destroy()方法是 HttpServlet 类中的方法。一个 Servlet 类可直接继承该方法，一般不需要重写。destroy()方法的声明格式如下：

```
public destroy()
```

当服务器终止服务时，如关闭服务器等，destroy()方法会被执行，消灭 Servlet 对象。

8.3 通过 JSP 页面调用 Servlet

尽管可以在浏览器的地址栏中直接输入 Servlet 对象的请求格式来运行一个 Servlet，但我们经常可能通过一个 JSP 页面来请求一个 Servlet。也就是说，我们可以让 JSP 页面负责数据的显示，而让一个 Servlet 去做与处理数据有关的事情。

本章中涉及的 JSP 页面存放在 Web 服务目录 chapter8 中，负责创建 Servlet 的字节码文件存放在 chapter8\WEB-INF\classes\star\moon 中。每当 Web 服务目录增加新的 Servlet 时，都需要为 web.xml 文件添加<servlet>标记和<servlet-mapping>标记，有关 web.xml 文件的配置请参见 8.1.2 节的内容。

1．通过表单向 Servlet 提交数据

任何一个 Web 服务目录中的 JSP 页面都可以通过表单或超链接请求某个 Servlet。通过 JSP 页面访问 Servlet 的好处是，JSP 页面可以负责页面的静态信息处理，动态信息处理交给 Servlet 去完成。

注意，如果 Servlet 的请求格式是"/name"，那么当前 Web 服务目录的 JSP 页面请求 Servlet 时必须写成"name"，不可以写成"/name"，否则将变成请求 root 服务目录中的某个 Servlet。

【例 8-1】 JSP 页面通过表单向 Servlet 提交一个正实数，Servlet 负责计算这个数的平方根返回给客户。web.xml 文件需要添加如下内容：

```
<servlet>
    <servlet-name>computer</servlet-name>
    <servlet-class>star.moon.Computer</servlet-class>
</servlet>
<servlet-mapping>
```

```
            <servlet-name>computer</servlet-name>
            <url-pattern>/getResult</url-pattern>
      </servlet-mapping>
```
JSP 页面如下：

givenumber.jsp（效果如图 8.4 所示）

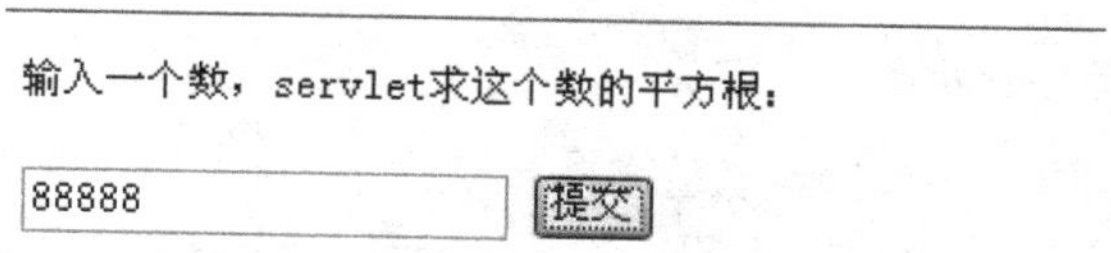

图 8.4　提交数字的 JSP 页面

```
<%@ page contentType="text/html;Charset=GB2312" %>
<HTML><BODY ><FONT size=2>
<P>输入一个数，servlet 求这个数的平方根：
<FORM action="getResult" method=post>
   <Input type=text name=number>
   <Input type=submit value="提交">
</FORM></BODY></HTML>
```

Servlet 源文件如下：

Computer.java（效果如图 8.5 所示）

图 8.5　负责计算平方根的 Servlet

```
package star.moon;
import java.io.*;
import javax.servlet.*;
import javax.servlet.http.*;
public class Computer extends HttpServlet{
    public void init(ServletConfig config) throws ServletException{
        super.init(config);
    }
    public void service(HttpServletRequest request,HttpServletResponse response)
                                                             throws IOException {
        response.setContentType("text/html;charset=GB2312");
        PrintWriter out=response.getWriter();
        out.println("<html><body>");
        String number=request.getParameter("number");      //获取客户提交的信息
        double n=0;
        try{  n=Double.parseDouble(number);
            out.print("<BR>"+number+"的平方根是：");
            out.print("<BR>"+Math.sqrt(n));
```

```
            }
            catch(NumberFormatException e) { out.print("<H1>请输入数字字符! </H1>"); }
            out.println("</body></html>");
        }
    }
```

2. 通过超链接访问 Servlet

我们可以在 JSP 页面中，单击一个超链接，访问 Servlet。下面的例 8-2 中的

【例 8-2】 Servlet 负责输出英文字母表。web.xml 文件需要添加如下内容：

```xml
<servlet>
    <servlet-name>show</servlet-name>
    <servlet-class>star.moon.ShowLetter</servlet-class>
</servlet>
<servlet-mapping>
    <servlet-name>show</servlet-name>
    <url-pattern>/helpMeShow</url-pattern>
</servlet-mapping>
```

JSP 页面如下：

showLetter.jsp

```jsp
<%@ page contentType="text/html;Charset=GB2312" %>
<HTML>
<BODY><FONT size=2>
<P><BR>单击超链接查看英文字母表：
    <BR><A href="helpMeShow">查看英文字母表</A>
</BODY>
</HTML>
```

Servlet 源文件如下：

ShowLetter.java

```java
package star.moon;
import java.io.*;
import javax.servlet.*;
import javax.servlet.http.*;
public class ShowLetter extends HttpServlet {
    public void init(ServletConfig config) throws ServletException {
        super.init(config);
    }
    public void service(HttpServletRequest request,HttpServletResponse response)
                                                        throws IOException {
        response.setContentType("text/html;charset=GB2312");
        PrintWriter out=response.getWriter();
        out.println("<html><body>");
        out.print("<BR>英文大写字母: ");
        for(char c='A';c<='Z';c++)
```

```
            out.print(" "+c);
        out.print("<BR>英文小写字母: ");
        for(char c='a';c<='z';c++)
            out.print(" "+c);
        out.println("</body></html>");
    }
}
```

8.4 Servlet 的共享变量

Servlet 类是 HttpServlet 的一个子类，那么在编写子类时就可以声明某些成员变量。当用户请求加载 Servlet 时，服务器分别为每个用户各启动一个线程，在该线程中，Servlet 调用 service()方法响应客户请求，那么 Servlet 类的成员变量是被所有线程共享的数据。

数学上有一个计算 π 的公式：

$$\frac{\pi}{4} = 1 - \frac{1}{3} + \frac{1}{5} - \frac{1}{7} + \frac{1}{9} - \frac{1}{11} \cdots$$

下面的例 8-3 利用成员变量被所有客户共享这一特性来计算 π 的值，即每当客户请求访问 Servlet 时都参与了一次 π 的计算。

客户通过单击一个 JSP 页面的超链接访问一个计算 π 的 Servlet。

web.xml 文件需要添加如下内容：

```
<servlet>
    <servlet-name>ok</servlet-name>
    <servlet-class>star.moon.computerPI</servlet-class>
</servlet>
<servlet-mapping>
    <servlet-name>ok</servlet-name>
    <url-pattern>/showPI</url-pattern>
</servlet-mapping>
```

【例 8-3】
JSP 页面如下：
showPI.jsp（效果如图 8.6 所示）
```
<%@ page contentType="text/html;charset=GB2312" %>
<HTML><BODY ><FONT size=5>
<A href="showPI" >查看 PI 的值</A>
</BODY></HTML>
```
Servlet 源文件如下：
ComputerPI.java（效果如图 8.7 所示）

查看PI的值

现在PI的值是：
PI= 3.1495925256000317

图 8.6　调用 Servlet 的页面 　　　　　　　图 8.7　负责计算 π 值的 Servlet

```java
package star.moon;
import java.io.*;
import javax.servlet.*;
import javax.servlet.http.*;
public class ComputerPI extends HttpServlet {
    double sum=0,i=1,j=1;
    int number=0;
    public void init(ServletConfig config) throws ServletException {
        super.init(config);
    }
    public synchronized void service(HttpServletRequest request,
                                     HttpServletResponse response) throws IOException {
        response.setContentType("text/html;charset=GB2312");
        PrintWriter out=response.getWriter();
        out.println("<html><body>");
        number++;
        sum=sum+i/j;
        j=j+2;
        i=-i;
        out.println("现在 PI 的值是:");
        out.println("<BR> PI= "+4*sum);
        out.println("</body></html>");
    }
}
```

8.5 doGet()方法和 doPost()方法

HttpServlet 类除了 init()、service()、destroy()方法外，还有两个很重要的方法：doGet()和 doPost()，用来处理客户的请求并做出响应。

当服务器创建 Servlet 对象后，该对象会调用 init()方法初始化自己，以后每当服务器再接收到 Servlet 请求时，就会产生一个新线程，在该线程中让 Servlet 对象调用 service()方法来检查 HTTP 请求类型（GET、POST 等），并在 service()方法中根据用户的请求方式，对应地再调用 doGet()或 doPost()方法。因此，在 Servlet 类中，我们不必重写 service()方法来响应客户，直接继承 service()方法即可。我们可以在 Servlet 类中重写 doPost()或 doGet()方法来响应用户的请求，这样可以增加响应的灵活性，并降低服务器的负担。

如果不论用户请求类型是 POST 还是 GET，服务器的处理过程完全相同，那么可以只在 doPost()方法中编写处理过程，而在 doGet()方法中再调用 doPost()方法即可，或只在 doGet()方法中编写处理过程，而在 doPost()方法中再调用 doGet()方法。如果根据请求的类型进行不同的处理，就需在两个方法中编写不同的处理过程。

在下面的例 8-6 中，用户可以使用表单向两个 Servlet 之一提交数字。一个 Servlet（由 GetSqare 负责创建）处理数据的手段不依赖表单提交数据的方式，无论是 POST 还是 GET，处理数据的手段相同，都是计算出表单提交的数字的平方；另一个 Servlet（由 GetSqareOrCubic 负责创建）处理数据的手段依赖表单提交数据的方式是 POST 还是 GET，提交方式是 POST 时，该 Servlet 计算表单提交的数的平方，提交方式是 GET 时，计算表单提交的数的立方。

web.xml 文件需要添加如下内容：
```xml
<servlet>
    <servlet-name>computer1</servlet-name>
    <servlet-class>star.moon.GetSquare</servlet-class>
</servlet>
<servlet>
    <servlet-name>computer2</servlet-name>
    <servlet-class>star.moon.GetSquareOrCubic</servlet-class>
</servlet>
<servlet-mapping>
    <servlet-name>computer1</servlet-name>
    <url-pattern>/showSquare</url-pattern>
</servlet-mapping>
<servlet-mapping>
    <servlet-name>computer2</servlet-name>
    <url-pattern>/showSquareOrCubic</url-pattern>
</servlet-mapping>
```

【例 8-4】

JSP 页面文件如下：

method.jsp
```jsp
<%@ page contentType="text/html;charset=GB2312" %>
<HTML><BODY ><FONT size=2>
<P>输入一个数，提交给 servlet（POST 方式）：
<FORM action="showSquare" method=post>
  <Input type=text name=number>
  <Input type=submit value="提交">
</FORM>
<P>输入一个数，提交给 servlet（GET 方式）：
<FORM action="showSquareOrCubic" method=get>
  <Input type=text name=number>
  <Input type=submit value="提交">
</FORM>
<P>输入一个数，提交给 servlet（POST 方式）：
<FORM action="showSquareOrCubic" method=get>
  <Input type=text name=number>
  <Input type=submit value="提交">
</FORM>
</BODY></HTML>
```

Servlet 源文件如下：

GetSqare.java
```java
package star.moon;
import java.io.*;
import javax.servlet.*;
```

```java
import javax.servlet.http.*;
public class GetSquare extends HttpServlet{
    public void init(ServletConfig config) throws ServletException{
        super.init(config);
    }
    public void doPost(HttpServletRequest request,HttpServletResponse response)
                                    throws ServletException,IOException {
        response.setContentType("text/html;charset=GB2312");
        PrintWriter out=response.getWriter();
        out.println("<html><body>");
        String number=request.getParameter("number");      //获取客户提交的信息
        double n=0;
        try{   n=Double.parseDouble(number);
               out.print("<BR>"+number+"的平方是：");
               out.print("<BR>"+n*n);
        }
        catch(NumberFormatException e) { out.print("<H1>请输入数字字符! </H1>"); }
        out.println("</body></html>");
    }
    public void doGet(HttpServletRequest request,HttpServletResponse response)
                                    throws  ServletException,IOException {
        doPost(request,response);
    }
}
```

GetSquareOrCubic.java

```java
package star.moon;
import java.io.*;
import javax.servlet.*;
import javax.servlet.http.*;
public class GetSquareOrCubic extends HttpServlet {
    public void init(ServletConfig config) throws ServletException {
        super.init(config);
    }
    public void doPost(HttpServletRequest request,HttpServletResponse response)
                                    throws ServletException,IOException {
        response.setContentType("text/html;charset=GB2312");
        PrintWriter out=response.getWriter();
        out.println("<html><body>");
        String number=request.getParameter("number");      //获取客户提交的信息
        double n=0;
        try{   n=Double.parseDouble(number);
               out.print("<BR>"+number+"的平方是：");
               out.print("<BR>"+n*n);
```

```
            }
                catch(NumberFormatException e) { out.print("<H1>请输入数字字符! </H1>"); }
                out.println("</body></html>");
        }
            public void doGet(HttpServletRequest request,HttpServletResponse response)
                                                        throws ServletException,IOException {
                response.setContentType("text/html;charset=GB2312");
                PrintWriter out=response.getWriter();
                out.println("<html><body>");
                String number=request.getParameter("number");        //获取客户提交的信息
                double n=0;
                try{  n=Double.parseDouble(number);
                    out.print("<BR>"+number+"的立方是： ");
                    out.print("<BR>"+n*n*n);
                }
                catch(NumberFormatException e) { out.print("<H1>请输入数字字符! </H1>"); }
                out.println("</body></html>");
        }
    }
```

8.6 重定向与转发

当用户请求一个 Servlet 时，Servlet 会调用 doPost()方法或 doGet()方法响应用户的请求。doPost()方法和 doGet()方法的两个参数的类型相同，都分别是 HttpServletRequest 和 HttpServletResponse，而且由服务器负责实例化。因此，在 Servlet 类中可以直接使用 doPost()方法或 doGet()方法中的参数。有关 HttpServletRequest 和 HttpServletResponse 类的用法在本书 2.1 节和 2.2 节都有详细介绍。本节学习在 Servlet 类中使用 HttpServletResponse 类的重定向方法 sendRedirect()，以及 RequestDispatcher 类的转发方法 forward()。

1. sendRedirect()方法

void sendRedirect(java.lang.String location)方法是 HttpServletResponse 类中的方法。当用户请求一个 Servlet 时，该 Servlet 在处理数据后，可以使用重定向方法 sendRedirect()方法将用户重新定向到一个 JSP 页面或另一个 Servlet。重定向方法仅仅是将用户从当前页面或 Servlet 定向到另一个 JSP 页面或 Servlet，不能将用户对当前页面的请求（HttpServletRequest 对象）转发给所定向的资源。

2. RequestDispatcher 对象

RequestDispatcher 对象可以把用户对当前 JSP 页面或 Servlet 的请求转发给另一个 JSP 页面或 Servlet，而且将用户对当前 JSP 页面或 Servlet 的请求和响应（HttpServletRequest 对象和 HttpServletResponse 对象）传递给所转发的 JSP 页面或 Servlet。

用户所请求的当前 JSP 或 Servlet 可以让 HttpServletRequest 对象调用

    public RequestDispatcher getRequestDispatcher(java.lang.String path)

返回一个 RequestDispatcher 对象，参数 path 是要转发的 JSP 页面或 Servlet 的地址。例如：

RequestDispatcher dispatcher= request.getRequestDispatcher("/a.jsp")

然后，RequestDispatcher 对象调用

void forward(ServletRequest request,ServletResponse response)

throws ServletException,ava.io.IOException

方法，可以将用户对当前 JSP 页面或 Servlet 的请求转发给 RequestDispatcher 对象所指定的 JSP 页面或 Servlet。例如：

dispatcher.forward (request, response);

将用户对当前 JSP 页面或 Servlet 的请求转变成对 a.jsp 页面的请求。

与重定向方法 sendRedirect()不同的是，用户在浏览器的地址栏中不能看到 forward()方法转发的页面地址或 Servlet 的地址，只能看到该页面或 Servlet 的运行效果；用户在浏览器的地址栏中看到的仍然是当前 Servlet 的地址（标记<Servlet-mapping>的访问格式）。

【例 8-5】 用户通过 input.jsp 页面提供的表单输入姓名和年龄，并提交给一个 Servlet（Verify 类负责创建）。如果用户输入的数据不完整（没有输入姓名或年龄）或输入的年龄不合法（如小于 1 或大于 150），那么该 Servlet 就将用户重新定向到 input.jsp 页面；如果用户输入的数据符合要求，Servlet 就将用户对 input.jsp 页面的请求转发到另一个 Servlet（ShowMessage 类负责创建），该 Servlet 将显示用户输入的信息。

web.xml 文件需要添加如下内容：

```
<servlet>
    <servlet-name>verify</servlet-name>
    <servlet-class>star.moon.Verify</servlet-class>
</servlet>
<servlet>
    <servlet-name>showMessage</servlet-name>
    <servlet-class>star.moon.ShowMessage</servlet-class>
</servlet>
<servlet-mapping>
    <servlet-name>verify</servlet-name>
    <url-pattern>/verifyYourMessage</url-pattern>
</servlet-mapping>
<servlet-mapping>
    <servlet-name>showMessage</servlet-name>
    <url-pattern>/forYouShowMessage</url-pattern>
</servlet-mapping>
```

JSP 页面文件如下：

input.jsp

```
<%@ page contentType="text/html;charset=GB2312" %>
<HTML><BODY ><FONT size=2>
<FORM action="verifyYourMessage" method=post>
    输入姓名： <Input type=text name=name>
    <BR>输入年龄： <Input type=text name=age>
    <BR><Input type=submit value="提交">
</FORM></BODY></HTML>
```

Servlet 源文件如下：
Verify.java
```java
package star.moon;
import java.io.*;
import javax.servlet.*;
import javax.servlet.http.*;
public class Verify extends HttpServlet {
    public void init(ServletConfig config) throws ServletException {
        super.init(config);
    }
    public void doPost(HttpServletRequest request,HttpServletResponse response)
                                        throws ServletException,IOException {
        String name=request.getParameter("name");      //获取客户提交的信息
        String age=request.getParameter("age");        //获取客户提交的信息
        int numberAge=Integer.parseInt(age);
        if(name.length()==0||name==null) {
            response.sendRedirect("input.jsp");         //重定向
        }
        else if(age.length()==0||name==null) {
            response.sendRedirect("input.jsp");         //重定向
        }
        else if(numberAge<=0||numberAge>=150) {
            response.sendRedirect("input.jsp");
        }
        else {
         RequestDispatcher dispatcher= request.getRequestDispatcher("forYouShowMessage");
            dispatcher.forward(request, response);      //重定向
        }
    }
    public void doGet(HttpServletRequest request,HttpServletResponse response)
                                         throws  ServletException,IOException {
        doPost(request,response);
    }
}
```
ShowMessage.java
```java
package star.moon;
import java.io.*;
import javax.servlet.*;
import javax.servlet.http.*;
public class ShowMessage extends HttpServlet {
    public void init(ServletConfig config) throws ServletException {
        super.init(config);
    }
```

```java
public void doPost(HttpServletRequest request,HttpServletResponse response)
                                      throws ServletException,IOException {
    response.setContentType("text/html;charset=GB2312");
    PrintWriter out=response.getWriter();
    String name=request.getParameter("name");         //获取客户提交的信息
    String age=request.getParameter("age");           //获取客户提交的信息
    try{   byte bb[]=name.getBytes("ISO-8859-1");
           name=new String(bb,"gb2312");
    }
    catch(Exception exp){ }
    out.print("<BR><Font color=blue size=4>您的姓名是：");
    out.print(name);
    out.print("<BR><Font color=pink size=2>您的年龄是：");
    out.print(age);
}
public void doGet(HttpServletRequest request,HttpServletResponse response)
                                      throws ServletException,IOException {
    doPost(request,response);
}
}
```

8.7 会话管理

HTTP 是一种无状态协议，一个客户向服务器发出请求（request）然后服务器返回响应（response），连接就被关闭了。在服务器端不保留连接的有关信息，因此当下一次连接时，服务器已没有以前的连接信息了，无法判断这一次连接与以前的连接是否属于同一客户。本节学习怎样在 Servlet 类中使用会话记录有关连接的信息，会话的原理与 4.3 节完全类似。

8.7.1 获取用户的会话

HttpServletRequest 对象 request 调用 getSession()方法获取用户的会话对象：

    HttpSession session=request.getSession(true);

一个用户在不同的 Servlet 中获取的 session 对象是完全相同的，不同用户的 session 对象互不相同。有关会话对象常用方法可参见 4.3 节。

【例 8-6】 编写两个 Servlet，当用户请求一个 Servlet 时，该 Servlet 将一个字符串对象存入用户的 session 对象中，然后用户请求另一个 Servlet，该 Servlet 将输出用户 session 对象中的字符串对象。

web.xml 文件需要添加如下内容：

```xml
<servlet>
    <servlet-name>boy</servlet-name>
    <servlet-class>star.moon.Boy</servlet-class>
</servlet>
<servlet>
```

```xml
            <servlet-name>look</servlet-name>
            <servlet-class>star.moon.Look</servlet-class>
        </servlet>
        <servlet-mapping>
            <servlet-name>boy</servlet-name>
            <url-pattern>/lookBoy</url-pattern>
        </servlet-mapping>
        <servlet-mapping>
            <servlet-name>look</servlet-name>
            <url-pattern>/lookMySession</url-pattern>
        </servlet-mapping>
```

Servlet 源文件如下：

Boy.java（效果如图 8.8 所示）

```java
package star.moon;
import java.io.*;
import javax.servlet.*;
import javax.servlet.http.*;
public class Boy extends HttpServlet {
    public void init(ServletConfig config) throws ServletException {
        super.init(config);
    }
    public void doPost(HttpServletRequest request,HttpServletResponse response)
                                            throws ServletException,IOException {
        response.setContentType("text/html;charset=GB2312");
        PrintWriter out=response.getWriter();
        out.print("<html><body>");
        HttpSession session=request.getSession(true);   //获取客户的会话对象
        session.setAttribute("name","耿祥义");
        out.println("您的会话的 id：");
        out.println("<BR>"+session.getId());
        out.println("<BR>单击超链接请求另一个 Servlet：");
        out.println("<A href=lookMySession>请求另一个 Servlet</A>");
        out.print("</body></html>");
    }
    public void doGet(HttpServletRequest request,HttpServletResponse response)
                                            throws ServletException,IOException {
        doPost(request,response);
    }
}
```

Look.java（效果如图 8.9 所示）

```java
package star.moon;
import java.io.*;
import javax.servlet.*;
import javax.servlet.http.*;
```

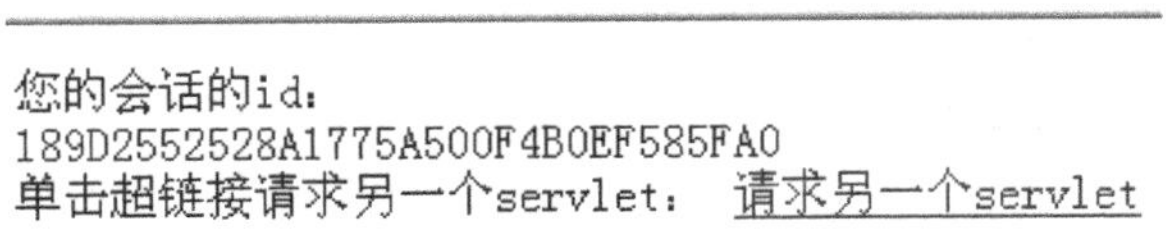

您的会话的id：
189D2552528A1775A500F4B0EF585FA0
单击超链接请求另一个servlet： 请求另一个servlet

图 8.8 获取会话

您的会话的id：
189D2552528A1775A500F4B0EF585FA0
您的会话中的数据：
耿祥义

图 8.9 获取会话中的数据

```java
public class Look extends HttpServlet {
    public void init(ServletConfig config) throws ServletException {
        super.init(config);
    }
    public void doPost(HttpServletRequest request,HttpServletResponse response)
                                    throws ServletException,IOException {
        response.setContentType("text/html;charset=GB2312");
        PrintWriter out=response.getWriter();
        out.print("<html><body>");
        HttpSession session=request.getSession(true);      //获取客户的会话对象
        String str=(String)session.getAttribute("name");   //获取会话中存储的数据
        out.println("您的会话的 id：");
        out.println("<BR>"+session.getId());
        out.println("<BR>您的会话中的数据：");
        out.println("<BR>"+str);
        out.print("</body></html>");
    }
    public void doGet(HttpServletRequest request,HttpServletResponse response)
                                    throws ServletException,IOException {
        doPost(request,response);
    }
}
```

8.7.2 猜数字

当客户访问或刷新 getNumber.jsp 页面时，随机分配给客户一个 1～100 之间的数，然后将这个数字存在用户的 session 对象中。然后，用户超链接到 inputNumber.jsp 页面，输入自己的猜测，并提交给一个 Servlet（HandleGuess 类负责创建），该 Servlet 负责处理用户的猜测，具体处理方式如下：

- 如果用户猜小了，就将用户重新定向到 inputNumber.jsp，并将"您猜小了"存放到用户的会话中。
- 如果用户猜大了，就将用户重新定向到 inputNumber.jsp，并将"您猜大了"存放到用户的会话中。
- 如果用户猜成功了，就将用户重新定向到 inputNumber.jsp，并将"您猜对了"存放到用户的会话中。

web.xml 文件需要添加如下内容：

```xml
<servlet>
    <servlet-name>guess</servlet-name>
    <servlet-class>star.moon.HandleGuess</servlet-class>
```

```xml
        </servlet>
        <servlet-mapping>
            <servlet-name>guess</servlet-name>
            <url-pattern>/handleGuess</url-pattern>
        </servlet-mapping>
```

【例 8-7】

JSP 页面文件如下：

getNumber.jsp（效果如图 8.10 所示）

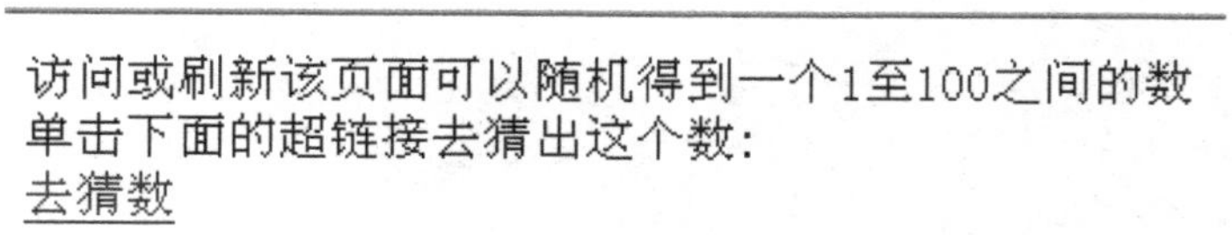

```
访问或刷新该页面可以随机得到一个1至100之间的数
单击下面的超链接去猜出这个数:
去猜数
```

图 8.10　随机获得一个 1~100 之间的数

```jsp
<%@ page contentType="text/html;charset=GB2312" %>
<HTML>
<BODY ><FONT size=4>
<% session.setAttribute("message","请您猜数");
   int randomNumber=(int)(Math.random()*100)+1;          //获取一个随机数
   session.setAttribute("savedNumber",new Integer(randomNumber));
%>
<P>访问或刷新该页面可以随机得到一个 1 至 100 之间的数
<BR>单击下面的超链接去猜出这个数:
<BR><A href="inputNumber.jsp">去猜数</A>
</BODY></HTML>
```

inputNumber.jsp（效果如图 8.11 所示）

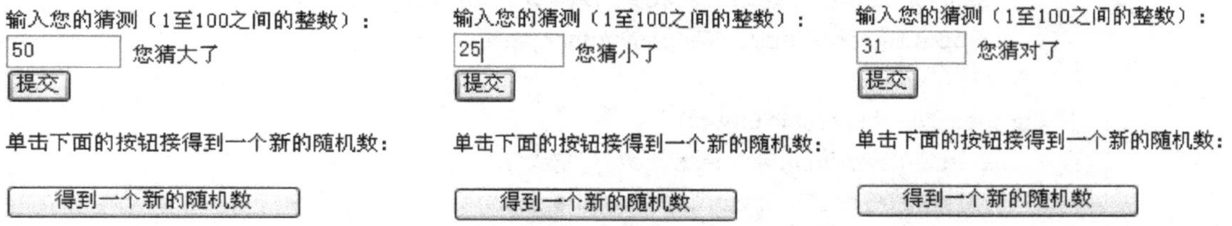

图 8.11　猜大小

```jsp
<%@ page contentType="text/html;charset=GB2312" %>
<HTML>
<BODY ><FONT size=2>
<%
    String message=(String)session.getAttribute("message");    //获取会话中的信息
%>
<FORM action="handleGuess" method=post>
    输入您的猜测（1 至 100 之间的整数）:
<BR><Input type=text name=clientGuessNumber size=8>
    <%= message%>
```

```html
            <BR><Input Type=submit value="提交">
        </FORM>
            单击下面的按钮得到一个新的随机数：
        <FORM action="getNumber.jsp" method=post>
            <Input type=submit value="得到一个新的随机数">
        </FORM>
    </BODY></HTML>
```

Servlet 源文件如下：

HandleGuess.java

```java
    package star.moon;
    import java.io.*;
    import javax.servlet.*;
    import javax.servlet.http.*;
    public class HandleGuess extends HttpServlet {
        public void init(ServletConfig config) throws ServletException{
            super.init(config);
        }
        public void doPost(HttpServletRequest request,HttpServletResponse response)
                                            throws ServletException,IOException {
            HttpSession session=request.getSession(true);   //获取客户的会话对象
            //获取客户猜测提交的数:
            int guessNumber=Integer.parseInt(request.getParameter("clientGuessNumber"));
            Integer integer=(Integer)session.getAttribute("savedNumber");//会话中存储的数
            int realNumber=integer.intValue();
            if(guessNumber<realNumber) {
                session.setAttribute("message","您猜小了");
                response.sendRedirect("inputNumber.jsp");
            }
            if(guessNumber>realNumber) {
                session.setAttribute("message","您猜大了");
                response.sendRedirect("inputNumber.jsp");
            }
            if(guessNumber==realNumber) {
                session.setAttribute("message","您猜对了");
                response.sendRedirect("inputNumber.jsp");
            }
        }
        public void doGet(HttpServletRequest request,HttpServletResponse response)
                                            throws ServletException,IOException {
            doPost(request,response);
        }
    }
```

习 题 8

1. Servlet 对象是在服务器端还是在客户端被创建？
2. Servlet 对象被创建后，将首先调用哪个方法？
3. Servlet 对象会在许多线程中调用哪个方法？
4. 假设创建 Servlet 的类是 tom.jiafei.Dalian，应当怎样配置 web.xml 文件？
5. HttpServletResponse 类的 sendRedirect()方法和 RequestDispatcher 类的 forward()方法有何不同？
6. Servlet 怎样获得用户的会话对象？

第 9 章

基于 Servlet 的 MVC 模式

> **本章导读**
> ✿ 知识点：理解 MVC 模式的核心思想——"视图"、"模型"和"控制器"；掌握 MVC 模式在 JSP 中的具体体现。
> ✿ 重点：掌握在 MVC 模式中怎样让 Servlet 用构造方法创建 JavaBean。
> ✿ 难点：掌握"控制器"的重要作用。
> ✿ 关键实践：根据某实际问题，用 MVC 模式设计一个 Web 应用。

JSP 页面擅长数据的显示，即适合作为用户的视图，应当尽量避免在 JSP 中使用大量的 Java 程序片来处理数据。Servlet 擅长数据的处理，应当尽量避免在 Servlet 中使用 out 流输出大量的 HTML 标记来显示数据。

一些小型的 Web 应用可以使用 JSP 页面调用 JavaBean 完成数据的处理，实现数据表示与数据处理的分离。在这种 JSP+JavaBean 模式中，JavaBean 不仅要提供修改和返回数据的方法，而且要经常参与数据的处理。当 Web 应用变得复杂时，我们希望 JavaBean 仅仅负责提供修改和返回数据的方法即可，不必参与数据的具体处理，而是把数据的处理交给称为控制器的 Servlet 对象去完成，即 Servlet 控制器负责处理数据，并将有关的结果存储到 JavaBean 中，实现存储与处理的分离。负责视图功能的 JSP 页面只要使用 JavaBean 标记显示 JavaBean 中的数据即可。

MVC 模式的核心思想是有效地组合"视图"、"模型"和"控制器"。本章将介绍 MVC 模式，掌握该模式对于设计合理的 Web 应用框架有着十分重要的意义。

本章使用的 Web 服务目录是 chapter9，chapter9 是在 Tomcat 安装目录的 webapps 目录下建立的 Web 服务目录。

现在需要在当前 Web 服务目录下建立目录结构 chapter9\WEB-INF\classes，然后根据 Servlet 的包名，在 classes 下再建立相应的子目录，如 Servlet 类的包名为 rain.snow，那么在 classes 下建立子目录 \rain\snow；如果 JavaBean 类的包名为 flower.grass，那么在 classes 下建立子目录 \flower\grass。为了让 Tomcat 服务器启用上述目录，必须重新启动 Tomcat 服务器。

9.1 MVC 模式介绍

模型-视图-控制器（Model-View-Controller，MVC）是一种先进的设计模式，是 Trygve Reenskaug 教授于 1978 年最早开发的一个设计模板或基本结构，其目的是以会话的形式提供方便的 GUI 支持。MVC 设计模式首先出现在 Smalltalk 编程语言中。

MVC 是一种通过 3 个不同部分构造一个软件或组件的理想办法：

- 模型（Model）——存储数据的对象。
- 视图（View）——为模型提供数据显示的对象。
- 控制器（Controller）——负责具体的业务逻辑操作，即控制器根据视图提出的要求对数据做出处理，并将有关结果存储到模型中，同时负责让模型和视图进行必要的交互，当模型中的数据变化时，让视图更新显示。

从面向对象的角度看，MVC 结构可以使程序更具有对象化特性，也更容易维护。在设计程序时，可以将某个对象看做"模型"，然后为"模型"提供恰当的显示组件，即"视图"。在 MVC 模式中，"视图"、"模型"和"控制器"之间是松耦合结构，便于系统的维护和扩展。

9.2 JSP 中的 MVC 模式

在 JSP 技术中，"视图"、"模型"和"控制器"的具体实现如下。

模型（Model）：一个或多个 JavaBean 对象，用于存储数据，JavaBean 主要提供简单的 setXXX()方法和 getXXX()方法，在这些方法中不涉及对数据的具体处理细节。

视图（View）：一个或多个 JSP 页面，为模型提供数据显示，JSP 页面主要使用 HTML 标记和 JavaBean 标记来显示数据。

控制器（Controller）：一个或多个 Servlet 对象，根据视图提交的要求进行数据处理操作，并将有关的结果存储到 JavaBean 中，然后 Servlet 使用重定向方式请求视图中的某个 JSP 页面更新显示，即让该 JSP 页面通过使用 JavaBean 标记显示控制器存储在 JavaBean 中的数据。

MVC 模式的结构如图 9.1 所示。

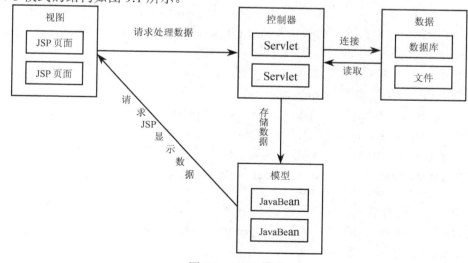

图 9.1 MVC 模式

9.3 模型的生命周期与视图更新

MVC 模式与前面学习的 JSP+JavaBean 模式有很大的不同。在 JSP+JavaBean 模式中，由 JSP 页面使用 JavaBean 标记

    `<jsp:useBean   id="给 bean 起的名字"   class="创建 bean 的类"   scope="bean 有效范围" />`

创建 JavaBean。在 MVC 模式中，由控制器 Servet 负责创建 JavaBean，并将有关数据存储到所创建的 JavaBean 中，然后 Servlet 请求某个 JSP 页面使用 JavaBean 的 getProperty 动作标记

    `<jsp:getProperty   name="bean 的名字"   property="bean 的属性" />`

来显示这个 JavaBean 的中的数据。

在 MVC 模式中，由 Servlet 负责用构造方法创建 JavaBean，因此创建 JavaBean 的类可以有带参数的构造方法，其他方法的命名继续保留"get"规则，但可以不遵守"set"规则。因为我们不希望 JSP 页面修改 JavaBean 中的数据，只需要它显示 JavaBean 中的数据。

在 MVC 模式中，Servlet 创建 JavaBean 也涉及生命周期，生命周期分为 request、session 和 application。以下假设创建 JavaBean 的类是 CreateJavaBeanClass，该类的包名为 flower.grass，并就 3 种情形分别讨论。

1. request 周期的 JavaBean

（1）JavaBean 的创建

Servlet 创建生命周期为 request 的 JavaBean 的步骤如下：

<1> 用 CreateJavaBeanClass 类的某个构造方法创建 Javabean 对象。例如：

    `CreateJavaBeanClass bean=new CreateJavaBeanClass();`

<2> 将所创建的 JavaBean 对象存放到 HttpServletRequest 对象 request 中，并指定查找该 JavaBean 的关键字。例如：

    `request.setAttribute("keyWord",bean);`

（2）视图更新

Servlet 创建 JavaBean 的第<2>步决定了 JavaBean 的生命周期为 request，该 JavaBean 只对 Servlet 请求的 JSP 页面有效。对于生命周期为 request 的 JavaBean，Servlet 使用 RequestDispatcher 对象向某个 JSP 页面发出请求，该 JSP 页面显示 JavaBean 中的数据，然后该 JavaBean 所占有的内存被释放、结束自己的生命。

Servlet 所请求的 JSP 页面，如 show.jsp 页面，必须使用如下标记获得 Servlet 所创建的 JavaBean 的引用（不负责创建 bean）：

    `<jsp:useBean id="keyWord" type="flower.grass. CreateJavaBeanClass" scope="request" />`

上述标记中的 id 就是 Servlet 所创建的 JavaBean。

Servlet 请求一个 JSP 页面文件 show.jsp 的代码如下：

    `RequestDispatcher dispatcher= request.getRequestDispatcher("/show.jsp");`
    `dispatcher.forward(request, response);`

如果上述代码执行成功，用户就看到了 show.jsp 页面的执行效果，它使用

    `<jsp:getProperty name="keyWord" property="Javabean 的变量" />`

标记显示 JavaBean 中的数据。

2. session 周期的 JavaBean

（1）JavaBean 的创建

Servlet 创建生命周期为 session 的 JavaBean 的步骤如下：

<1> 用 CreateJavabeanClass 类的某个构造方法创建 JavaBean 对象，例如：

  CreateJavaBeanClass bean=new CreateJavaBeanClass();

<2> 将所创建的 JavaBean 对象存放到 HttpServletSession 对象 session 中，并指定查找该 JavaBean 的关键字。例如：

  HttpSession session=request.getSession(true);
  session.setAttribute("keyWord",bean);

（2）视图更新

Servlet 创建 JavaBean 的第<2>步决定了 JavaBean 的生命周期为 session，该 JavaBean 的有效期限为用户的会话期间。当 Servlet 创建生命周期为 session 的 JavaBean 后，只要用户的会话没有消失，该 JavaBean 就一直存在。

一个用户在访问 Web 服务目录的各个 JSP 中都可以使用

  <jsp:useBean id="keyWord" type="flower.grass.CreateJavabeanClass" scope="session" />

标记获得 Servlet 所创建的 JavaBean 的引用，然后使用

  <jsp:getProperty name="keyWord" property="JavaBean 的变量" />

标记显示该 JavaBean 中的数据。

对于生命周期为 session 的 JavaBean，如果 Servlet 希望某个 JSP 显示其中的数据，可以使用 RequestDispatcher 对象向该 JSP 页面发出请求，也可以使用 HttpServletResponse 类中的重定向方法 sendRedirect()。

注意： 不同用户的 session 生命周期的 JavaBean 是互不相同的，即占有不同的内存空间。

3. application 周期的 JavaBean

（1）JavaBean 的创建

Servlet 创建生命周期为 application 的 JavaBean 的步骤如下：

<1> 用 CreateJavaBeanClass 类的某个构造方法创建 JavaBean 对象。例如：

  CreateJavaBeanClass bean=new CreateJavaBeanClass();

<2> Servlet 可以使用 getServletContext()方法返回服务器创建的 ServletContext 对象的引用，将所创建的 JavaBean 对象存放到服务器创建的 ServletContext 对象中，并指定查找该 JavaBean 的关键字。例如：

  getServletContext().setAttribute("keyWord",bean);

（2）视图更新

Servelt 创建 JavaBean 的第<2>步决定了 JavaBean 的生命周期为 application。当 Servlet 创建生命周期为 application 的 JavaBean 后，只要 Web 应用程序不结束，该 JavaBean 就一直存在。一个用户在访问 Web 服务目录的各 JSP 中都可以使用

  <jsp:useBean id="keyWord" type="flower.grass.CreateJavaBeanClass" scope="application" />

标记获得 Servlet 所创建的 JavaBean 的引用，然后使用

  <jsp:getProperty name="keyWord" property="Javabean 的变量" />

标记显示该 JavaBean 中的数据。

对于生命周期为 application 的 JavaBean，如果 Servlet 希望某个 JSP 显示其中的数据，可以使用 RequestDispatcher 对象向该 JSP 页面发出请求，也可以使用 HttpServletResponse 类中的重定向方法 sendRedirect()。

注意：所有用户的 application 生命周期的 JavaBean 是相同的，即占有相同的内存空间。

9.4 MVC 模式的简单实例

以下结合几个简单的例子体现 MVC 三部分的设计和实现。按照本章约定，下述 JavaBean 类的包名均为 flower.grass，Servlet 类的包名均为 rain.snow。需要将编译通过的 JavaBean 类和 Servlet 类的字节码件分别复制到目录 chapter9\WEB-INF\classes\flower\grass 和 chapter9\WEB-INF\classes\rain\snow 中。

9.4.1 计算三角形的面积

现在设计一个 Web 应用，该 Web 应用提供两个 JSP 页面，一个页面使得用户可以输入三角形三条边的长度；另一个页面可以显示三角形三条边的长度和面积。Web 应用提供一个 Servlet，负责处理三角形三条边数据，并根据三条边计算三角形的面积，然后将有关数据存储到 Javabean 中；Web 应用提供的 JavaBean 负责刻画三角形模型，提供简单的获取数据和修改数据的方法。

【例 9-1】编写 web.xml 文件并保存到 Web 服务目录的子目录 WEB-INF 中，即 chapter9\WEB-INF 中。web.xml 文件的内容如下（需要用纯文本编辑，保存格式为 UTF-8 编码）：

web.xml
```xml
<?xml version="1.0" ?>
<web-app>
    <servlet>
        <servlet-name>handle</servlet-name>
        <servlet-class>star.moon.HandleData</servlet-class>
    </servlet>
    <servlet-mapping>
        <servlet-name>handle</servlet-name>
        <url-pattern>/handleData</url-pattern>
    </servlet-mapping>
</web-app>
```

① 模型（JavaBean）

由于 Servlet 类中要使用 JavaBean，所以为了能顺利地编译 Servlet 类和 JavaBean 类，将下面的 Trangle.java 保存到目录 D:\rain\snow\flower\grass 中。进入上述目录，编译 Triangle.java，并将编译后的字节码文件 Triangle.class 复制到目录 chapter9\WEB-INF\classes\flower\grass 中。

Triangle.java
```java
package flower.grass;
public class Triangle{
    double sideA,sideB,sideC;
```

```
        double area;
        boolean isTriangle;
        public void setSideA(double a) {
            sideA=a;
        }
        public double getSideA(){
            return sideA;
        }
        public void setSideB(double b) {
            sideB=b;
        }
        public double getSideB(){
            return sideB;
        }
        public void setSideC(double c) {
            sideC=c;
        }
        public double getSideC(){
            return sideC;
        }
        public void setArea(double area) {
            this.area=area;
        }
        public double getArea(){
            return area;
        }
        public void setIsTriangle(boolean boo) {
            isTriangle=boo;
        }
        public boolean getIsTriangle(){
            return isTriangle;
        }
    }
```

注意： 在该 JavaBean 中，getArea()方法并没有参与面积的计算，只是简单返回存储的数据 area，这与 5.4.1 节中的 JavaBean 有所不同。

② 视图（JSP 页面）

input.jsp（效果如图 9.2 所示）

```
    <%@ page contentType="text/html;Charset=GB2312" %>
    <HTML><BODY ><FONT size=2>
    <FORM action="handleData" method="post" >
      <P>输入三角形的三条边（提交给 Servlet 去处理）：
      <BR>边 A：<Input type=text name="sideA" value=0 size=4>
           边 B：<Input type=text name="sideB" value=0 size=4>
           边 C：<Input type=text name="sideC" value=0 size=4>
      <Input type=submit value="提交">
    </FORM>
    </BODY></HTML>
```

show.jsp（效果如图 9.3 所示）

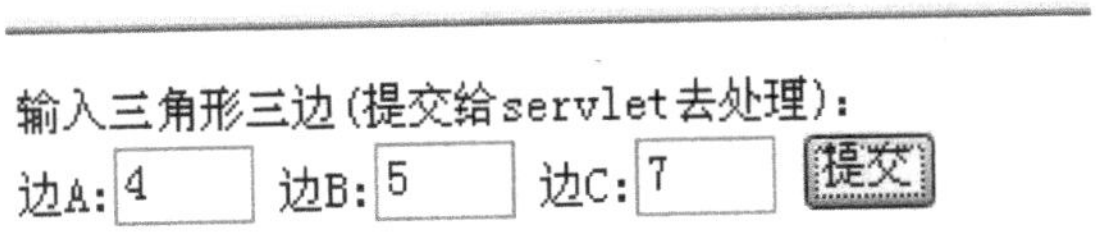

图 9.2　输入三角形三条边数据 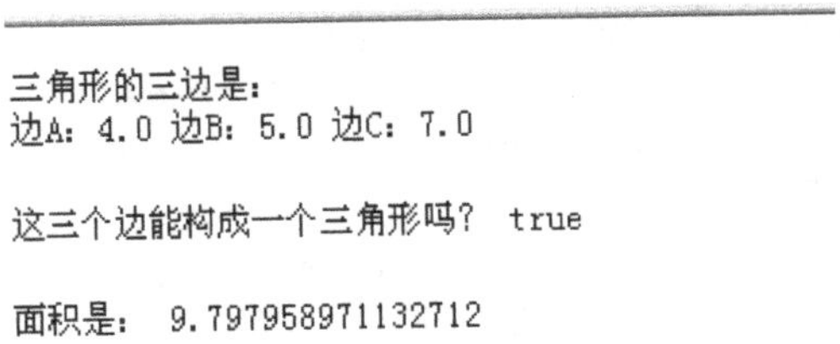　　　图 9.3　显示数据

```
<%@ page contentType="text/html;Charset=GB2312" %>
<%@ page import="flower.grass.Triangle"%>
<jsp:useBean id="triangle" type="flower.grass.Triangle" scope="request" />
<HTML><BODY ><Font size=2>
<P>三角形的三条边是：
<BR>边 A：<jsp:getProperty   name="triangle"   property="sideA" />
     边 B：<jsp:getProperty   name="triangle"   property="sideB" />
     边 C：<jsp:getProperty   name="triangle"   property="sideC" />
<P>这三条边能构成一个三角形吗?
<jsp:getProperty   name= "triangle"   property="isTriangle" />
<P>面积是：
   <jsp:getProperty   name= "triangle"   property="area" />
</FONT>
</BODY></HTML>
```

③ 控制器（Servlet）

　　Servlet 类中要使用 JavaBean，为了能顺利编译 Servlet 类和 JavaBean 类，将下面的 HandleData.java 保存到目录 D:\rain\snow 中。进入上述目录，编译 HandeData.java，并将编译得到的字节码文件 HandeData.class 复制到 chapter9\WEB-INF\classes\rain\snow 中。

HandleData.java

```
    package rain.snow;
    import flower.grass.*;
    import java.io.*;
    import javax.servlet.*;
    import javax.servlet.http.*;
    public class HandleData extends HttpServlet{
        public void init(ServletConfig config) throws ServletException{
             super.init(config);
        }
        public void doPost(HttpServletRequest request,HttpServletResponse response)
                                                throws ServletException,IOException {
             Triangle tri=new Triangle();           //创建 Javabean 对象
             request.setAttribute("triangle",tri);  //将 tri 存储到 HttpServletRequest 对象中
             double a=Double.parseDouble(request.getParameter("sideA"));
             double b=Double.parseDouble(request.getParameter("sideB"));
             double c=Double.parseDouble(request.getParameter("sideC"));
             double p=(a+b+c)/2.0;
             double area=Math.sqrt(p*(p-a)*(p-b)*(p-c));
```

```
            tri.setSideA(a);                  //将数据存储在 tri 中
            tri.setSideB(b);                  //将数据存储在 tri 中
            tri.setSideC(c);                  //将数据存储在 tri 中
            tri.setArea(area);                //将数据存储在 tri 中
            if(a+b>c&&b+c>a&&a+c>b)
                tri.setIsTriangle(true);
            else
                tri.setIsTriangle(false);
            RequestDispatcher dispatcher= request.getRequestDispatcher("/show.jsp");
            dispatcher.forward(request, response);    //请求 show.jsp 显示 tri 中的数据
        }
        public void doGet(HttpServletRequest request,HttpServletResponse response)
                                                throws ServletException,IOException {
            doPost(request,response);
        }
    }
```

9.4.2 四则运算

设计一个 Web 应用,有两个 JSP 页面(inputNumber.jsp 和 showResult.jsp)、一个 JavaBean 和一个 Servlet。JSP 页面 inputNumber.jsp 提供一个表单,用户可以通过表单输入两个数和四则运算符号提交给 Servlet 控制器;JavaBean 负责存储运算数、运算符号和运算结果。Servlet 控制器负责四则运算,将结果存储在 JavaBean 中,并负责请求 JSP 页面 showResult.jsp 显示 JavaBean 中的数据。

【例 9-2】 在 Servlet 转发的页面 showResult.jsp 中,用户可以继续输入运算数。

web.xml 文件需要添加如下内容:

```xml
<servlet>
    <servlet-name>handleComputer</servlet-name>
    <servlet-class>rain.snow.HandleComputer</servlet-class>
</servlet>
<servlet-mapping>
    <servlet-name>handleComputer </servlet-name>
    <url-pattern>/helpComputer</url-pattern>
</servlet-mapping>
```

① **模型(JavaBean)**

将下面的 ComputerBean.java 保存到目录 D:\rain\snow\flower\grass 中。进入上述目录,编译 ComputerBean.java,并将编译得到的字节码文件 ComputerBean.class 复制到目录 chapter9\WEB-INF\classes\flower\grass 中。

ComputerBean.java
```java
        package flower.grass;
        public class ComputerBean {
            double numberOne,numberTwo,result;
            String operator="+";
            public void setNumberOne(double n) {
```

```
            numberOne=n;
        }
        public double getNumberOne(){
            return numberOne;
        }
        public void setNumberTwo(double n) {
            numberTwo=n;
        }
        public double getNumberTwo(){
            return numberTwo;
        }
        public void setOperator(String s) {
            operator=s.trim();
        }
        public String getOperator(){
            return operator;
        }
        public void setResult(double r) {
            result=r;
        }
        public double getResult(){
            return result;
        }
    }
```

请读者比较上述 JavaBean 与 5.4.2 节中的不同。

② **视图（JSP 页面）**

inputNumber.jsp（效果如图 9.4 所示）

```
<%@ page contentType="text/html;charset=GB2312" %>
<%@ page import="flower.grass.ComputerBean" %>
<HTML><BODY ><FONT size=2>
  <FORM action="helpComputer" method=post name=form>
   <TABLE>
    <TR><TD> 输入两个数:</TD>
        <TD> <Input type=text name="numberOne" value=0 size=6></TD>
        <TD> <Input type=text name="numberTwo" value=0 size=6></TD>
    </TR>
    <TR>
       <TD>选择运算符号:</TD>
       <TD> <Select name="operator">
             <Option value="+">+(加)
             <Option value="-">-（减）
             <Option value="*">*（乘）
             <Option value="/">/（除）
           </Select>
       </TD>
       <TD> <Input type="submit" value="提交选择" name="submit"></TD>
    </TR>
```

```
        </TABLE>
        </FORM>
    </BODY></HTML>
```
showResult.jsp（效果如图 9.5 所示）

图 9.4　输入运算数、选择运算符号

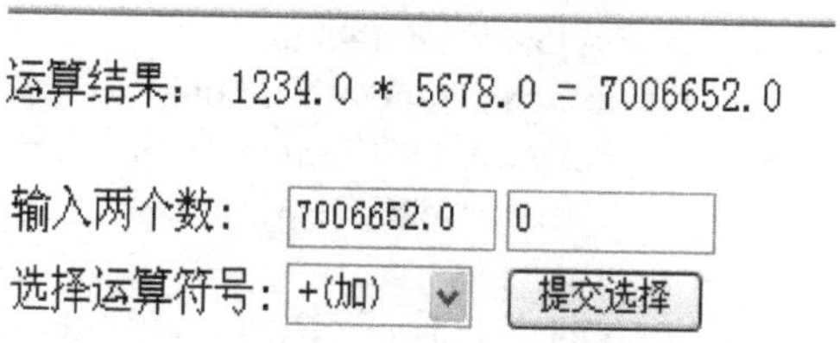

图 9.5　显示结果与继续计算

```
<%@ page contentType="text/html;charset=GB2312" %>
<%@ page import="flower.grass.ComputerBean" %>
<HTML><BODY ><FONT size=3>
  <jsp:useBean id="ok" type="flower.grass.ComputerBean" scope="session"/>
      运算结果：
  <jsp:getProperty name="ok" property="numberOne"/>
  <jsp:getProperty name="ok" property="operator"/>
  <jsp:getProperty name="ok" property="numberTwo"/> =
  <jsp:getProperty name="ok" property="result"/>
    <FORM action="helpComputer" method=post name=form>
    <TABLE>
    <TR><TD> 输入两个数：</TD>
        <TD> <Input type=text name="numberOne" value="<jsp:getProperty
                  name="ok" property="result"/>"size=10></TD>
        <TD> <Input type=text name="numberTwo" value=0 size=10></TD>
    </TR>
    <TR>
        <TD>选择运算符号：</td>
        <TD> <Select name="operator">
              <Option value="+">+(加)
              <Option value="-">-（减）
              <Option value="*">*（乘）
              <Option value="/">/（除)
            </Select>
        </TD>
        <TD> <Input type="submit" value="提交选择" name="submit"></TD>
    </TR>
    </TABLE>
  </BODY></HTML>
```

② **控制器（Servlet）**

由于 Servlet 类中要使用 JavaBean，所以为了能顺利编译 Servlet 类和 JavaBean 类，将 HandleComputer.java 保存到目录 D:\rain\snow 中。进入上述目录，编译 HandeComputer.java，并将编译后的字节码文件 HandeComputer.class 复制到目录 chapter9\WEB-INF\classes\rain\snow 中。

HandleComputer.java

```java
package rain.snow;
import flower.grass.*;
import java.io.*;
import javax.servlet.*;
import javax.servlet.http.*;
public class HandleComputer extends HttpServlet{
    public void init(ServletConfig config) throws ServletException{
        super.init(config);
    }
    public void doPost(HttpServletRequest request,HttpServletResponse response)
                                            throws ServletException,IOException {
        ComputerBean dataBean=null;
        HttpSession session=request.getSession(true);
        try{   dataBean=(ComputerBean)session.getAttribute("ok");
            if(dataBean==null) {
                dataBean=new ComputerBean();        //创建 Javabean 对象
                session.setAttribute("ok",dataBean); //将 dataBean 存储到 session 对象中
            }
        }
        catch(Exception exp) {
            dataBean=new ComputerBean();          //创建 Javabean 对象
            session.setAttribute("ok",dataBean);   //将 dataBean 存储到 session 对象中
        }
        double numberOne=Double.parseDouble(request.getParameter("numberOne"));
        double numberTwo=Double.parseDouble(request.getParameter("numberTwo"));
        String operator=request.getParameter("operator");
        double result=0;
        if(operator.equals("+")){
            result=numberOne+numberTwo;
        }
        else if(operator.equals("-")){
            result=numberOne-numberTwo;
        }
        else if(operator.equals("*")){
            result=numberOne*numberTwo;
        }
        else if(operator.equals("/")){
            result=numberOne/numberTwo;
        }
        dataBean.setNumberOne(numberOne);     //将数据存储在 dataBean 中
        dataBean.setNumberTwo(numberTwo);     //将数据存储在 dataBean 中
        dataBean.setOperator(operator);        //将数据存储在 dataBean 中
        dataBean.setResult(result);            //将数据存储在 dataBean 中
        RequestDispatcher dispatcher= request.getRequestDispatcher("/showResult.jsp");
        dispatcher.forward(request, response);   //请求 showResult.jsp 显示 dataBean 中的数据
    }
    public void doGet(HttpServletRequest request,HttpServletResponse response)
```

```
            doPost(request,response);                    throws ServletException,IOException {
        }
    }
```

9.4.3 读取文件

【例 9-3】 设计一个 Web 应用，有两个 JSP 页面（choiceFile.jsp 和 showFile.jsp）、一个 JavaBean 和一个 Servlet。用户在 JSP 页面 choiceFile.jsp 中选择一个文件，提交给 Servlet，该 Servlet 负责读取文件的有关信息存放到 JavaBean 中，并请求 JSP 页面 showFile.jsp 显示 JavaBean 中的数据。

web.xml 文件需要添加如下内容：

```xml
<servlet>
    <servlet-name>handleFile</servlet-name>
    <servlet-class>rain.snow.HandleFile</servlet-class>
</servlet>
<servlet-mapping>
    <servlet-name>handleFile </servlet-name>
    <url-pattern>/helpReadFile</url-pattern>
</servlet-mapping>
```

① 模型（JavaBean）

将下面的 FileMessage.java 保存到目录 D:\rain\snow\flower\grass 中。进入上述目录，编译 FileMessage.java，将编译得到的字节码文件 FileMessage.class 复制到目录 chapter9\WEB-INF\classes\flower\grass 中。

FileMessage.java
```java
package flower.grass;
public class FileMessage {
    String filePath,fileName,fileContent;
    long fileLength;
    public void setFilePath(String str) {
        filePath=str;
    }
    public String getFilePath(){
        return filePath;
    }
    public void setFileName(String str) {
        fileName=str;
    }
    public String getFileName(){
        return fileName;
    }
    public void setFileContent(String str) {
        fileContent=str;
    }
    public String getFileContent(){
```

```
            return fileContent;
        }
        public void setFileLength(long len) {
            fileLength=len;
        }
        public long getFileLength(){
            return fileLength;
        }
    }
```

② 视图（JSP 页面）

choiceFile.jsp（效果如图 9.6 所示）

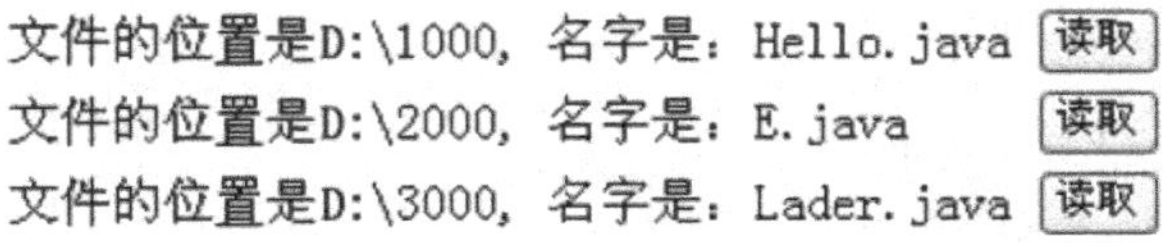

　　文件的位置是D:\1000，名字是：Hello.java [读取]
　　文件的位置是D:\2000，名字是：E.java   [读取]
　　文件的位置是D:\3000，名字是：Lader.java [读取]

图 9.6 选择文件

```
<%@ page contentType="text/html;Charset=GB2312" %>
<HTML><BODY><FONT size=3>
  <TABLE>
  <FORM action="helpReadFile" method="post" name="form">
    <TR>
     <TD>文件的位置是 D:\1000,<TD>
         <Input type="hidden" value="D:\\1000" name="filePath">
      <TD>名字是：Hello.java<TD>
         <Input type="hidden" value="Hello.java" name="fileName">
      <TD><Input type="submit" value="读取" name="submit">
    </TR>
  </FORM>
  <FORM action="helpReadFile" method="post" name="form">
    <TR>
     <TD>文件的位置是 D:\2000,<TD>
         <Input type="hidden" value="d:\\2000" name="filePath">
      <TD>名字是：E.java<TD>
         <Input type="hidden" value="E.java" name="fileName">
      <TD><Input type="submit" value="读取" name="submit">
    </TR>
  </FORM>
  <FORM action="helpReadFile" method="post" name="form">
    <TR>
     <TD>文件的位置是 D:\3000,<td>
         <Input type="hidden" value="D:\\3000" name="filePath">
      <TD>名字是：Lader.java<TD>
         <Input type="hidden" value="Lader.java" name="fileName">
      <TD><Input type="submit" value="读取" name="submit">
    </TR>
  </FORM>
```

</BODY></HTML>

showFile.jsp（效果如图 9.7 所示）

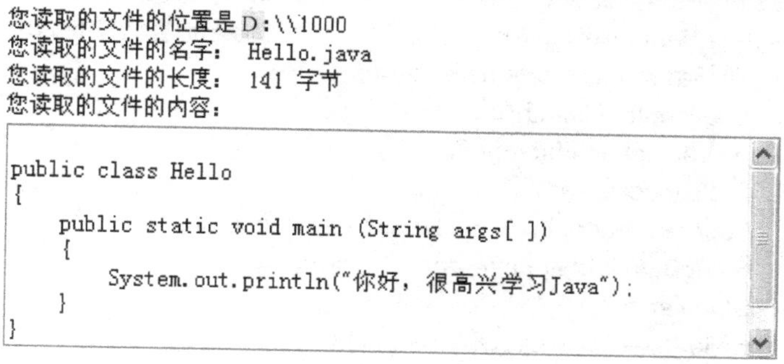

图 9.7 显示文件信息

```
<%@ page contentType="text/html;charset=GB2312" %>
<%@ page import="flower.grass.FileMessage" %>
 <jsp:useBean id="file" type="flower.grass.FileMessage" scope="request"/>
<HTML><BODY ><FONT size=2>
    您读取的文件的位置是
  <jsp:getProperty name="file" property="filePath"/>
  <BR> 您读取的文件的名字：
  <jsp:getProperty name="file" property="fileName"/>
  <BR> 您读取的文件的长度：
  <jsp:getProperty name="file" property="fileLength"/> 字节
  <BR> 您读取的文件的内容：
  <BR><TextArea  rows= "8" cols= "60">
      <jsp:getProperty name="file" property="fileContent" />
  </TextArea>
</BODY></HTML>
```

③ 控制器（**Servlet**）

将下面的 HandleFile.java 保存到目录 D:\rain\snow 中。进入上述目录，编译 HandeFile.java，并将编译后的字节码文件 HandeFile.class 复制到 chapter9\WEB-INF\classes\rain\snow 中。

HandleFile.java

```
package rain.snow;
import flower.grass.*;
import java.io.*;
import javax.servlet.*;
import javax.servlet.http.*;
public class HandleFile extends HttpServlet{
   public void init(ServletConfig config) throws ServletException{
      super.init(config);
   }
   public void doPost(HttpServletRequest request,HttpServletResponse response)
                                       throws ServletException,IOException {
      FileMessage file=new FileMessage();         //创建 Javabean 对象
      request.setAttribute("file",file);          //将 file 存储到 request 对象中
```

```
            String filePath=request.getParameter("filePath");
            String fileName=request.getParameter("fileName");
            file.setFilePath(filePath);              //将数据存储在 file 中
            file.setFileName(fileName);              //将数据存储在 file 中
            try{ File f=new File(filePath,fileName);
                 long length=f.length();
                 file.setFileLength(length);
                 FileReader in=new FileReader(f) ;
                 BufferedReader inTwo=new BufferedReader(in);
                 StringBuffer stringbuffer=new StringBuffer();
                 String s=null;
                 while ((s=inTwo.readLine())!=null) {
                     stringbuffer.append("\n"+s);
                 }
                 String content=new String(stringbuffer);
                 file.setFileContent(content);       //将数据存储在 file 中
            }
            catch(IOException exp) { }
            RequestDispatcher dispatcher= request.getRequestDispatcher("/showFile.jsp");
            dispatcher.forward(request, response);   //请求 showFile.jsp 显示 file 中的数据
        }
        public void doGet(HttpServletRequest request,HttpServletResponse response)
                                               throws ServletException,IOException {
            doPost(request,response);
        }
    }
```

9.4.4 查询数据库

【例 9-4】 设计一个 Web 应用,有两个 JSP 页面(choiceDatabase.jsp 和 showRecord.jsp)、一个 JavaBean 和一个 Servlet。用户在 JSP 页面 choiceDatabase.jsp 中选择一个数据库中的表,提交给 Servlet,该 Servlet 负责分页读取表中的记录,并把读取的记录存放到 JavaBean 中,然后请求 JSP 页面 showRecord.jsp 显示 JavaBean 中的数据。showRecord.jsp 页面提供了【上一页】、和【下一页】按钮,用户在该页面可以继续请求控制器 Servelt,以便读取上一页或下一页。

本 Web 应用使用了 7.7 节中的 CachedRowSetImpl 对象。将数据库查询结果保存到 ResultSet 对象后,就可以关闭和数据库的连接。

web.xml 文件需要添加如下内容:

```xml
<servlet>
    <servlet-name>database</servlet-name>
    <servlet-class>rain.snow.HandleDatabase</servlet-class>
</servlet>
<servlet-mapping>
    <servlet-name>database </servlet-name>
    <url-pattern>/helpReadRecord</url-pattern>
</servlet-mapping>
```

① 模型（JavaBean）

将下面的 ShowRecordByPage.java 文件保存到目录 D:\rain\snow\flower\grass 中。进入上述目录，编译 ShowRecordByPage.java 文件，将编译后的字节码文件 ShowRecordByPage.class 复制到目录 chapter9\WEB-INF\classes\flower\grass 中。

ShowRecordByPage.java

```java
package flower.grass;
import com.sun.rowset.*;
public class ShowRecordByPage{
    CachedRowSetImpl rowSet=null;              //存储表中全部记录的行集对象
    int pageSize=10;                           //每页显示的记录数
    int pageAllCount=0;                        //分页后的总页数
    int showPage=1;                            //当前显示页
    StringBuffer presentPageResult;            //显示当前页内容
    String databaseName="";                    //数据库名称
    String tableName="";                       //表的名字
    StringBuffer formTitle=null;               //表头
    public void setRowSet(CachedRowSetImpl set) {
        rowSet=set;
    }
    public CachedRowSetImpl getRowSet(){
        return rowSet;
    }
    public void setPageSize(int size) {
        pageSize=size;
    }
    public int getPageSize(){
        return pageSize;
    }
    public int getPageAllCount(){
        return pageAllCount;
    }
    public void setPageAllCount(int n) {
        pageAllCount=n;
    }
    public void setShowPage(int n) {
        showPage=n;
    }
    public int getShowPage(){
        return showPage;
    }
    public void setPresentPageResult(StringBuffer p) {
        presentPageResult=p;
    }
    public StringBuffer getPresentPageResult(){
        return presentPageResult;
    }
```

```java
        public void setDatabaseName(String s) {
            databaseName=s.trim();
        }
        public String getDatabaseName(){
            return databaseName;
        }
        public void setTableName(String s) {
            tableName=s.trim();
        }
        public String getTableName(){
            return tableName;
        }
        public void setFormTitle(StringBuffer s) {
            formTitle=s;
        }
        public StringBuffer getFormTitle(){
            return formTitle;
        }
    }
```

② 视图（JSP 页面）

choiceDatabase.jsp（效果如图 9.8 所示）

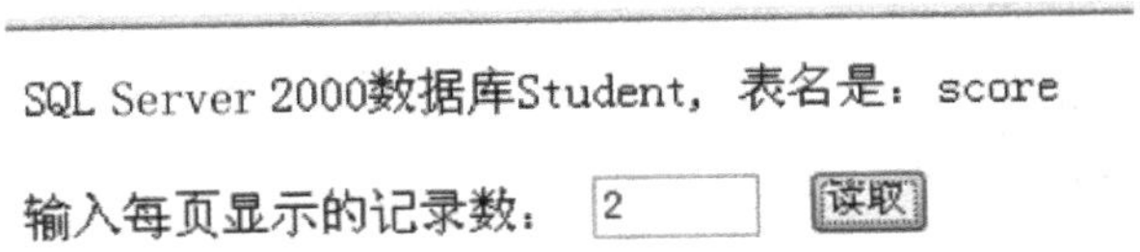

图 9.8　选择数据库和表

```jsp
<%@ page contentType="text/html;Charset=GB2312" %>
<HTML><BODY><FONT size=3>
    <TABLE>
    <FORM action="helpReadRecord" method="post" name="form">
    SQL Server 2000 数据库 Student，表名是：score
    <Input type="hidden" value="Student" name="databaseName">
    <Br>输入每页显示的记录数：
        <Input type="text" value="2" name="pageSize" size=6>
        <Input type="hidden" value="score" name="tableName">
        <Input type="submit" value="读取" name="submit">
    </FORM>
</BODY></HTML>
```

showRecord.jsp（效果如图 9.9 所示）

```jsp
<%@ page contentType="text/html;charset=GB2312" %>
<%@ page import="flower.grass.ShowRecordByPage" %>
<HTML><BODY><FONT size=3>
    <jsp:useBean id="database" type="flower.grass.ShowRecordByPage" scope="session"/>
    您查询的数据库：<jsp:getProperty name="database" property="databaseName"/>,
    查询的表：<jsp:getProperty name="database" property="tableName"/>。
```

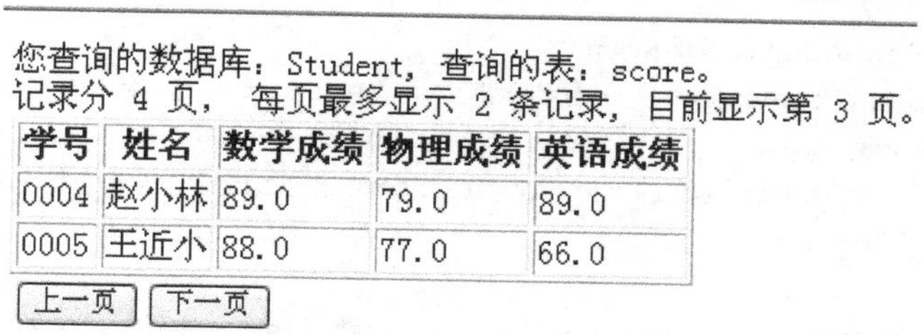

图 9.9 分页显示记录

```
  <BR>记录分 <jsp:getProperty name="database" property="pageAllCount"/> 页,
    每页最多显示 <jsp:getProperty name="database" property="pageSize"/> 条记录,
    目前显示第 <jsp:getProperty name="database" property="showPage"/> 页。
<TABLE border=1>
  <jsp:getProperty name="database" property="formTitle"/>
  <jsp:getProperty name="database" property="presentPageResult"/>
</TABLE>
<TABLE>
  <TR><TD>
      <FORM action="helpReadRecord" method="post" name="form">
        <Input type="hidden" value="previousPage" name="whichPage">
        <Input type="submit" value="上一页" name="submit">
      </FORM>
    </TD>
    <TD>
      <FORM action="helpReadRecord" method="post" name="form">
        <Input type="hidden" value="nextPage" name="whichPage">
        <Input type="submit" value="下一页" name="submit">
      </FORM>
    </TD>
  </TR>
</FORM>
</BODY></HTML>
```

③ **控制器（Servlet）**

将下面的 HandleDatabase.java 文件保存到目录 D:\rain\snow 中。进入上述目录，编译 HandleDatabase.java 文件，并将编译后的字节码文件 HandleDatabase.class 复制到目录 chapter9\WEB-INF\classes\rain\snow 中。

HandleDatabase.java
```
    package rain.snow;
    import flower.grass.*;
    import com.sun.rowset.*;
    import java.sql.*;
    import java.io.*;
    import javax.servlet.*;
    import javax.servlet.http.*;
    public class HandleDatabase extends HttpServlet{
       int 字段个数;
```

```java
            CachedRowSetImpl rowSet=null;
            public void init(ServletConfig config) throws ServletException{
                super.init(config);
                try { Class.forName("com.microsoft.sqlserver.jdbc.SQLServerDriver"); }
                catch(Exception e){ }
            }
            public void doPost(HttpServletRequest request,HttpServletResponse response)
                                                                throws ServletException,IOException {
                Connection con;
                StringBuffer presentPageResult=new StringBuffer();
                ShowRecordByPage databaseBean=null;
                HttpSession session=request.getSession(true);
                try{  databaseBean=(ShowRecordByPage)session.getAttribute("database");
                      if(databaseBean==null) {
                          databaseBean=new ShowRecordByPage();       //创建 Javabean 对象
                          session.setAttribute("database",databaseBean);
                      }
                }
                catch(Exception exp) {
                     databaseBean=new ShowRecordByPage();
                     session.setAttribute("database",databaseBean);
                }
                String databaseName=request.getParameter("databaseName");
                String tableName=request.getParameter("tableName");
                String ps= request.getParameter("pageSize");
                if(ps!=null) {
                    try{  int mm=Integer.parseInt(ps);
                          databaseBean.setPageSize(mm);
                    }
                    catch(NumberFormatException exp) { databaseBean.setPageSize(1); }
                }
                int showPage=databaseBean.getShowPage();
                int pageSize=databaseBean.getPageSize();
                boolean boo=databaseName!=null&&tableName!=null&&
                        databaseName.length()>0&&tableName.length()>0;
                if(boo) {
                   databaseBean.setDatabaseName(databaseName);     //数据存储在 databaseBean 中
                   databaseBean.setTableName(tableName);           //数据存储在 databaseBean 中
                   String uri="jdbc:sqlserver://127.0.0.1:1433;DatabaseName="+databaseName;
                   try{  字段个数=0;
                        con=DriverManager.getConnection(uri,"sa","sa");
                        DatabaseMetaData metadata=con.getMetaData();
                        ResultSet rs1=metadata.getColumns(null,null,tableName,null);
                        int k=0;
                        String 字段[]=new String[100];
                        while(rs1.next()){
                            字段个数++;
```

```
            字段[k]=rs1.getString(4);            //获取字段的名字
            k++;
        }
        StringBuffer str=new StringBuffer();
        str.append("<tr>");
        for(int i=0;i<字段个数;i++) {
            str.append("<th>"+字段[i]+"</th>");
        }
        str.append("</tr>");
        databaseBean.setFormTitle(str);           //数据存储在 databaseBean 中
        Statement sql=con.createStatement(ResultSet.TYPE_SCROLL_SENSITIVE,
                                 ResultSet.CONCUR_READ_ONLY);
        ResultSet rs=sql.executeQuery("SELECT * FROM "+tableName);
        rowSet=new CachedRowSetImpl();            //创建行集对象
        rowSet.populate(rs);
        con.close();                              //关闭连接
        databaseBean.setRowSet(rowSet);           //数据存储在 databaseBean 中
        rowSet.last();
        int m=rowSet.getRow();                    //总行数
        int n=pageSize;
        int pageAllCount=((m%n)==0)?(m/n):(m/n+1);
        databaseBean.setPageAllCount(pageAllCount);  //数据存储在 databaseBean 中
    }
    catch(SQLException exp) { }
}
String whichPage=request.getParameter("whichPage");
if(whichPage==null||whichPage.length()==0) {
    showPage=1;
    databaseBean.setShowPage(showPage);
    CachedRowSetImpl rowSet=databaseBean.getRowSet();
    if(rowSet!=null) {
        presentPageResult=show(showPage,pageSize,rowSet);
        databaseBean.setPresentPageResult(presentPageResult);
    }
}
else if(whichPage.equals("nextPage")) {
    showPage++;
    if(showPage>databaseBean.getPageAllCount())
        showPage=1;
    databaseBean.setShowPage(showPage);
    CachedRowSetImpl rowSet=databaseBean.getRowSet();
    if(rowSet!=null) {
        presentPageResult=show(showPage,pageSize,rowSet);
        databaseBean.setPresentPageResult(presentPageResult);
    }
}
else if(whichPage.equals("previousPage")) {
```

```
                    showPage--;
                    if(showPage<=0)
                        showPage=databaseBean.getPageAllCount();
                    databaseBean.setShowPage(showPage);
                    CachedRowSetImpl rowSet=databaseBean.getRowSet();
                    if(rowSet!=null) {
                        presentPageResult=show(showPage,pageSize,rowSet);
                        databaseBean.setPresentPageResult(presentPageResult);
                    }
                }
                databaseBean.setPresentPageResult(presentPageResult);
                RequestDispatcher dispatcher= request.getRequestDispatcher("/showRecord.jsp");
                dispatcher.forward(request, response);        //请求 showRecord.jsp 显示数据
            }
            public StringBuffer show(int page,int pageSize,CachedRowSetImpl rowSet) {
                StringBuffer str=new StringBuffer();
                try{  rowSet.absolute((page-1)*pageSize+1);
                    for(int i=1;i<=pageSize;i++) {
                        str.append("<tr>");
                        for(int k=1;k<=字段个数;k++) {
                            str.append("<td>"+rowSet.getString(k)+"</td>");
                        }
                        str.append("</tr>");
                        rowSet.next();
                    }
                }
                catch(SQLException exp) { }
                return str;
            }
            public void doGet(HttpServletRequest request, HttpServletResponse response)
                                                    throws ServletException,IOException {
                doPost(request,response);
            }
        }
```

习 题 9

1. 在 JSP 中，MVC 模式中的数据模型之角色由谁担当？
2. 在 JSP 中，MVC 模式中的控制器之角色由谁担当？
3. 在 JSP 中，MVC 模式中的视图之角色由谁担当？
4. MVC 的好处是什么？
5. MVC 模式中用到的 JavaBean 是由 JSP 页面还是 Servlet 负责创建？
6. 使用 MVC 模式设计一个猜数游戏。

第 10 章

实训一：会员管理系统

> **本章导读**
> ✪ 知识点：掌握 Web 应用中常见基本模块的开发方法。
> ✪ 重点：掌握 MVC 模式在设计 Web 应用中的重要性。
> ✪ 难点：掌握"登录"和"上传照片"模块的设计
> ✪ 关键实践：对"会员管理系统"进行调试。

本章讲述如何用 JSP 技术建立一个简单的会员管理系统，其目的是掌握一般 Web 应用中常用基本模块的开发方法。JSP 引擎为 Tomcat 5.5；系统采用 MVC 模式实现各个模块；数据库连接操作使用加载 Java 纯驱动程序方式，使得本系统可以采用任何提供 Java 纯驱动程序的数据库，本系统在实际应用中采用的是 SQL Server 2000 数据库。

10.1 系统模块构成

会员管理系统具有如下模块。

- **会员注册**：新会员填写表单，包括会员名、E-mail 地址等信息。如果输入的会员名已经被其他用户注册使用，系统提示新用户更改自己的会员名。
- **会员登录**：输入会员名、密码。如果用户输入的会员名或密码有错误，系统将显示错误信息。
- **上传照片**：如果登录成功，用户可以使用该模块上传照片。
- **浏览会员**：成功登录的会员可以分页浏览其他会员，如果用户直接进入该页面或没有成功登录就进入该页面，将被链接到"会员登录"页面。
- **修改密码**：成功登录的会员可以在该页面修改自己的登录密码，如果用户直接进入该页面或没有成功登录就进入该页面，将被链接到"会员登录"页面。
- **修改注册信息**：成功登录的会员可以在该页面修改自己的注册信息，如联系电话、通信地址等，如果用户直接进入该页面或没有成功登录就进入该页面，将被链接到"会员登录"页面。
- **退出登录**：成功登录的用户可以使用该模块退出登录。

10.2 数据库设计

系统采用加载纯 Java 数据库驱动程序方式访问数据库。使用 SQL Server 2000 建立一个数据库 ComeHere，该数据库共有一个表，表的结构如下。

1. member 表的字段

会员的注册信息存入 member 表中，member 表的主键是 logname，各字段值的说明如下：

- logname　　存储会员登录名字。
- password　　存储会员登录密码。
- sex　　存储会员性别。
- age　　存储会员的年龄。
- phone　　存储会员的电话。
- email　　存储会员的 E-mail 地址。
- message　　存储会员的简历。
- pic　　存储会员照片的名字。

2. member 表的结构

member 表的详细结构设计如图 10.1 所示。

设计表“member”，位置是“ComeHere”中、“(local)”上

列名	数据类型	长度	允许空
logname	nvarchar	50	
password	char	10	
sex	char	10	
age	int	4	
phone	nvarchar	50	✓
email	nvarchar	50	✓
message	nvarchar	50	✓
pic	nvarchar	50	✓

图 10.1　member 表的结构

10.3 系统管理

本系统使用的 Web 服务目录是 chapter10。chapter10 是在 Tomcat 安装目录的 webapps 目录下建立的 Web 服务目录。

在当前 Web 服务目录下建立录 chapter10\WEB-INF\classes，然后根据 Servlet 的包名，在 classes 下再建立相应的子目录，如 Servlet 类的包名为 myservlet.control，那么在 classes 下建立子目录 \myservlet\control；如果 JavaBean 类的包名为 mybean.data，那么在 classes 下建立子目录 \mybean\data。为了让 Tomcat 服务器启用上述目录，必须重新启动 Tomcat 服务器。

1. 页面管理

本系统所用的 JSP 页面全部保存在 Web 服务目录 chapter10 中。

所有的页面将包括一个导航条，该导航条由注册、登录、上传照片、浏览会员、查看会员、留言板、修改密码、修改个人信息组成。为了减少代码的编写，其他页面通过使用 JSP 的<%@ include ...%>标记将导航条文件 head.txt 嵌入自己的页面，head.txt 保存在目录 chapter10 中。head.txt 的内容如下：

head.txt

```
<%@ page contentType="text/html;charset=GB2312" %>
<CENTER><FONT size=6 ><P>会 员 管 理 系 统</FONT></CENTER>
<TABLE   cellSpacing="1" cellPadding="1" width="500" align="center" border="0" >
  <TR valign="bottom">
    <TD align="center"><A href="register.jsp"><B>会员注册</B></A></TD>
    <TD align="center"><A href="login.jsp"><B>会员登录</B></A></TD>
    <TD align="center"><A href="upload.jsp"><B>上传照片</B></A></TD>
    <TD align="center"><A href="choiceLookType.jsp"><B>浏览会员</B></A></TD>
  </TR>
  <TR valign="bottom">
    <TD align="center"><A href="choiceModifyMess.jsp"><B>修改注册信息</B></A></TD>
    <TD align="center"><A href="modifyPassword.jsp"><B>修改密码</B></A></TD>
    <TD align="center"><A href="helpExitLogin"><B>退出登录</B></A></TD>
    <TD align="center"><A href="index.jsp"><B>返回主页</B></A></TD>
  </TR>
</FONT>
</TABLE>
```

主页 index.jsp 由导航条、一个欢迎语和一幅图片 welcome.jpg 组成，welcome.jpg 保存在目录 chapter10 中。

index.jsp

```
<%@ page contentType="text/html;charset=GB2312" %>
<HTML> <BODY>
<HEAD><%@ include file="head.txt" %></HEAD>
<CENTER> <H1><FONT size=9 color=red>欢迎您的到来</FONT></H1>
<IMG src="welcome.jpg" width=800 height=800 ></IMG>
</CENTER></H1>
</BODY></HTML>
```

用户可以通过在浏览器的地址栏中输入 "http://服务器 IP:8080/index.jsp" 或 "http://服务器 IP:8080/" 访问该主页，主页运行效果如图 10.2 所示。

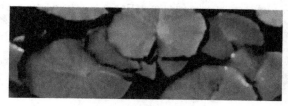

图 10.2 主页 index.jsp

2. JavaBean 与 Servlet 管理

本系统的 JavaBean 类的包名均为 mybean.data；Servlet 类的包名均为 myservlet.control。需要将编译通过的 JavaBean 类和 Servlet 类的字节码件分别复制到目录 chapter10\WEB-INF\classes\mybean\data 和 chapter10\WEB-INF\classes\myservlet\control 中。

由于 Servlet 类中要使用 JavaBean，所以为了能顺利编译 Servlet 类和 JavaBean 类，需要将 JavaBean 类和 Servlet 类分别保存到 D:\myservlet\control\mybean\data 和 D:\myservlet\control 目录中。分别进入上述目录，编译 JavaBean 类和 Servlet 类。

3．配置文件管理

本系统的 Servlet 类的包名均为 myservlet.control，需要配置 Web 服务目录的 web.xml 文件。根据本书使用的 Tomcat 安装目录及使用的 Web 服务目录，将下面的 web.xml 文件保存到 D:\apache-tomcat-5.5.20\webapps\chapter10\WEB-INF 中（有关 web.xml 文件的配置规定可参见 8.1.2 节）。

web.xml

```xml
<?xml version="1.0" encoding="ISO-8859-1" ?>
<web-app>
<servlet>
    <servlet-name>register</servlet-name>
    <servlet-class>myservlet.control.HandleRegister</servlet-class>
</servlet>
<servlet>
    <servlet-name>login</servlet-name>
    <servlet-class>myservlet.control.HandleLogin</servlet-class>
</servlet>
<servlet>
    <servlet-name>upload</servlet-name>
    <servlet-class>myservlet.control.HandleUpload</servlet-class>
</servlet>
<servlet>
    <servlet-name>showImage</servlet-name>
    <servlet-class>myservlet.control.ShowImage</servlet-class>
</servlet>
<servlet>
    <servlet-name>lookRecord</servlet-name>
    <servlet-class>myservlet.control.HandleDatabase</servlet-class>
</servlet>
<servlet>
    <servlet-name>modifyPassword</servlet-name>
    <servlet-class>myservlet.control.HandlePassword</servlet-class>
</servlet>
<servlet>
    <servlet-name>getOldMess</servlet-name>
    <servlet-class>myservlet.control.GetOldMess</servlet-class>
</servlet>
<servlet>
    <servlet-name>modifyOldMess</servlet-name>
    <servlet-class>myservlet.control.HandleModifyMess</servlet-class>
</servlet>
```

```xml
<servlet>
    <servlet-name>exit</servlet-name>
    <servlet-class>myservlet.control.HandleExit</servlet-class>
</servlet>
<servlet-mapping>
    <servlet-name>register</servlet-name>
    <url-pattern>/helpRegister</url-pattern>
</servlet-mapping>
<servlet-mapping>
    <servlet-name>login</servlet-name>
    <url-pattern>/helpLogin</url-pattern>
</servlet-mapping>
<servlet-mapping>
    <servlet-name>upload</servlet-name>
    <url-pattern>/helpUpload</url-pattern>
</servlet-mapping>
<servlet-mapping>
    <servlet-name>showImage</servlet-name>
    <url-pattern>/helpShowImage</url-pattern>
</servlet-mapping>
<servlet-mapping>
    <servlet-name>lookRecord</servlet-name>
    <url-pattern>/helpShowMember</url-pattern>
</servlet-mapping>
<servlet-mapping>
    <servlet-name>modifyPassword</servlet-name>
    <url-pattern>/helpModifyPassword</url-pattern>
</servlet-mapping>
<servlet-mapping>
    <servlet-name>getOldMess</servlet-name>
    <url-pattern>/helpGetOldMess</url-pattern>
</servlet-mapping>
<servlet-mapping>
    <servlet-name>modifyOldMess</servlet-name>
    <url-pattern>/helpModifyMess</url-pattern>
</servlet-mapping>
<servlet-mapping>
    <servlet-name>exit</servlet-name>
    <url-pattern>/helpExitLogin</url-pattern>
</servlet-mapping>
</web-app>
```

10.4 会员注册

当新会员注册时，该模块要求用户必须输入会员名、密码信息，否则不允许注册。用户的注册信息被存入数据库的 member 表中。

该模块的模型 JavaBean 描述用户的注册信息；视图部分由两个 JSP 页面构成，一个 JSP 页面负责提交用户的注册信息到控制器，另一个 JSP 页面负责显示注册是否成功的信息；控制器 Servlet 负责将视图提交的信息写入数据库的 member 表中，并负责更新视图。

1. 模型（JavaBean）

下列 JavaBean 的实例用来描述用户注册信息。

Register.java

```java
package mybean.data;
public class Register{
    String logname="",password="",sex="",
           email="",   phone="", message="";
    String backNews;
    int age;
    public void setLogname(String name) {
        logname=name;
    }
    public String getLogname(){
        return logname;
    }
    public void setAge(int n) {
        age=n;
    }
    public int getAge(){
        return age;
    }
    public void setSex(String s) {
        sex=s;
    }
    public String getSex(){
        return sex;
    }
    public void setPassword(String pw) {
        password=pw;
    }
    public String getPassword(){
      eturn password;
    }
    public void setEmail(String em) {
        email=em;
    }
    public String getEmail(){
        return email;
    }
    public void setPhone(String ph) {
        phone=ph;
    }
    public String getPhone(){
        return phone;
    }
```

```
    public String getMessage(){
        return message;
    }
    public void setMessage(String m) {
        message=m;
    }
    public String getBackNews(){
        return backNews;
    }
    public void setBackNews(String s) {
        backNews=s;
    }
}
```

2．视图（JSP 页面）

本模块视图有两个 JSP 页面：Register.jsp 和 showRegisterMess.jsp。Register.jsp 页面负责提供输入注册信息界面；showRegisterMess.jsp 负责显示注册反馈信息，如注册是否成功等。

Register.jsp（效果如图 10.3 所示）

图 10.3　填写注册信息

```
<%@ page contentType="text/html;charset=GB2312" %>
<HTML><HEAD><%@ include file="head.txt" %></HEAD>
<BODY><FONT size=2>
<CENTER>
<FORM action="helpRegister" name=form>
<BR>输入您的信息，会员名字必须由字母和数字组成，带*号项必须填写。
<TABLE>
    <TR><TD>会员名称:</TD> <TD><Input type=text name="logname" >*</TD></TR>
    <TR><TD>设置密码:</TD><TD><Input type=password name="password">*</TD></TR>
    <TR><TD>性别:</TD>
        <TD><Input type=radio   name="sex" checked="o" value="男">男
            <Input type=radio   name="sex" value="女">女
```

```
          </TD>
        </TR>
       <TR><TD>会员年龄:</td><td><Input type=text name="age" value="0"></TD></TR>
       <TR><TD>电子邮件:</td><td><Input type=text name="email"></TD></TR>
       <TR><TD>联系电话:</td><td><Input type=text name="phone"></TD></TR>
     </TABLE>
     <TABLE>
       <TR><TD>输入您的个人简介：</TD></TR>
       <TR><TD><TextArea name="message" Rows="6" Cols="30"></TextArea></TD> </TR>
       <TR><TD><Input type=submit name="g" value="提交"></TD> </TR>
     </TABLE>
  </FORM></CENTER>
</BODY></HTML>
```

showRegisterMess.jsp（效果如图 10.4 所示）

图 10.4 显示注册结果

```
<%@ page contentType="text/html;charset=GB2312" %>
<%@ page import="mybean.data.Register"%>
<jsp:useBean id="register" type="mybean.data.Register" scope="request" />
<HTML><HEAD><%@ include file="head.txt" %></HEAD>
<HTML><BODY bgcolor=cyan >
<CENTER><FONT size=4 color=blue >
   <BR> <jsp:getProperty name="register"   property="backNews"   />
   </FONT>
<FONT size=2>
<TABLE>
   <TR><TD>注册的会员名称:</td><td><jsp:getProperty name="register"
                                                 property="logname" /></TD></TR>
   <TR><TD>注册的性别:</TD> <TD><jsp:getProperty name="register" property="sex" /></TD></TR>
   <TR><TD>注册的会员年龄:</TD><TD><jsp:getProperty name="register"
                                                 property="age" /></TD></TR>
   <TR><TD>注册的电子邮件:</td><td><jsp:getProperty name="register"
                                                 property="email" /></TD></TR>
```

```
        <TR><TD>注册的联系电话:</td><td><jsp:getProperty name="register"
                                        property="phone" /></TD></TR>
    </TABLE>
    <TABLE><TR><TD>您输入的个人简介： </TD></TR>
        <TR><TD><TextArea name="message" Rows="6" Cols="30">
                <jsp:getProperty name="register" property="message" /></TextArea></TD> </TR>
    </TABLE>
    </FONT>
    </CENTER>
    </BODY></HTML>
```

3．控制器（Servlet）

Servlet 负责连接数据库，将用户提交的信息写入到 member 表，并将用户转发到 showRegisterMess.jsp 页面查看注册反馈信息。

HandleRegister.java

```java
package myservlet.control;
import mybean.data.*;
import java.sql.*;
import java.io.*;
import javax.servlet.*;
import javax.servlet.http.*;
public class HandleRegister extends HttpServlet {
    public void init(ServletConfig config) throws ServletException{
        super.init(config);
        try { Class.forName("com.microsoft.sqlserver.jdbc.SQLServerDriver"); }
        catch(Exception e) { }
    }
    public String handleString(String s) {
        try{   byte bb[]=s.getBytes("iso-8859-1");
            s=new String(bb);
        }
        catch(Exception ee) { }
        return s;
    }
    public void doPost(HttpServletRequest request,HttpServletResponse response)
                                            throws ServletException,IOException {
        Connection con;
        PreparedStatement sql;
        Register reg=new Register();
        request.setAttribute("register",reg);
        String logname=request.getParameter("logname").trim(),
            password=request.getParameter("password").trim(),
            sex=request.getParameter("sex").trim(),
            email=request.getParameter("email").trim(),
            phone=request.getParameter("phone").trim(),
            message=request.getParameter("message");
        int age=Integer.parseInt(request.getParameter("age").trim());
```

```java
String uri="jdbc:sqlserver://127.0.0.1:1433;DatabaseName=ComeHere";
if(logname==null)
    logname="";
if(password==null)
    password="";
boolean isLD=true;
for(int i=0;i<logname.length();i++) {
    char c=logname.charAt(i);
    if(!(((c<='z'&&c>='a')||(c<='Z'&&c>='A')||(c<='9'&&c>='0'))))
        isLD=false;
}
boolean boo=logname.length()>0&&password.length()>0&&isLD;
String backNews="";
try{   con=DriverManager.getConnection(uri,"sa","sa");
       String insertCondition="INSERT INTO member VALUES (?,?,?,?,?,?,?,?)";
       sql=con.prepareStatement(insertCondition);
       if(boo) {
          sql.setString(1,handleString(logname));
          sql.setString(2,handleString(password));
          sql.setString(3,handleString(sex));
          sql.setInt(4,age);
          sql.setString(5,phone);
          sql.setString(6,email);
          sql.setString(7,handleString(message));
          sql.setString(8,"public.jpg");
          int m=sql.executeUpdate();
          if(m!=0) {
              backNews="注册成功";
              reg.setBackNews(backNews);
              reg.setLogname(logname);
              reg.setPassword(handleString(password));
              reg.setAge(age);
              reg.setSex(handleString(sex));
              reg.setEmail(handleString(email));
              reg.setPhone(phone);
              reg.setMessage(handleString(message));
          }
       }
       else {
          backNews="信息填写不完整或名字中有非法字符";
          reg.setBackNews(backNews);
       }
       con.close();
}
catch(SQLException exp) {
    backNews="该会员名已被使用，请您更换名字"+exp;
    reg.setBackNews(backNews);
```

```
                }
                RequestDispatcher dispatcher= request.getRequestDispatcher("/showRegiterMess.jsp");
                    dispatcher.forward(request, response);
            }
            public void doGet(HttpServletRequest request,HttpServletResponse response)
                                                throws ServletException,IOException {
                doPost(request,response);
            }
        }
```

10.5 会员登录

用户可在该模块输入自己的会员名和密码，系统将对会员名和密码进行验证，如果输入用户名或密码有错误，将提示用户输入的用户名或密码不正确。

该模块的模型 JavaBean 描述用户登录的信息；视图部分由两个 JSP 页面构成，一个 JSP 页面负责提交用户的登录信息到控制器，另一个 JSP 页面负责显示登录是否成功的信息；控制器 Servlet 负责验证会员名和密码是否正确，并负责更新视图。

1．模型（JavaBean）

下列 JavaBean 的实例用来描述用户登录信息。
Login.java
```
        package mybean.data;
        public class Login {
            String logname,
                    password,
                    backNews="";
            boolean success=false;
            public void setLogname(String name) {
                logname=name;
            }
            public String getLogname(){
                return logname;
            }
            public void setPassword(String pw) {
                password=pw;
            }
            public String getPassword() {
                return password;
            }
            public String getBackNews() {
                return backNews;
            }
            public void setBackNews(String s) {
                backNews=s;
            }
            public void setSuccess(boolean b) {
```

```
            success=b;
        }
        public boolean getSuccess(){
            return success;
        }
    }
```

2．视图（JSP 页面）

本模块视图有两个 JSP 页面：login.jsp 和 showLoginMess.jsp。login.jsp 页面负责提供输入登录信息界面；showLoginMess.jsp 负责显示登录反馈信息，如登录是否成功等。

login.jsp（效果如图 10.5 所示）

图 10.5 输入登录信息

```
<%@ page contentType="text/html;charset=GB2312" %>
<HTML><HEAD><%@ include file="head.txt" %></HEAD>
<BODY bgcolor=pink><FONT size=2><CENTER>
<P>请您登录
<FORM action="helpLogin" Method="post">
    <BR>登录名称:<Input type=text name="logname">
    <BR>输入密码:<Input type=password name="password">
    <BR><Input type=submit name="g" value="提交">
</FORM></CENTER>
</BODY></HTML>
```

showLoginMess.jsp（效果如图 10.6 所示）

图 10.6 显示登录结果

```
<%@ page contentType="text/html;charset=GB2312" %>
<%@ page import="mybean.data.Login"%>
<jsp:useBean id="login" type="mybean.data.Login" scope="session" />
<HTML><HEAD><%@ include file="head.txt" %></HEAD>
<HTML><BODY bgcolor=yellow >
```

```
<CENTER><FONT size=4 color=blue >
    <BR> <jsp:getProperty name="login"  property="backNews"  />
    </FONT>
<%   if(login.getSuccess()==true) {
%>       <BR>登录会员名称:<jsp:getProperty name="login" property="logname" />
<%   }
      else {
%>       <BR>登录会员名称:<jsp:getProperty name="login" property="logname" />
         <BR>登录会员密码:<jsp:getProperty name="login" property="password" />
<%   }
%>
</FONT></CENTER>
</BODY></HTML>
```

3. 控制器（Servlet）

控制器负责连接数据库，查询 member 表，验证用户输入的会员名和密码是否在 member 表中，并将用户转发到 showRegisterMess.jsp 页面查看登录反馈信息。

HandleLogin.java

```java
package myservlet.control;
import mybean.data.*;
import java.sql.*;
import java.io.*;
import javax.servlet.*;
import javax.servlet.http.*;
public class HandleLogin extends HttpServlet {
    public void init(ServletConfig config) throws ServletException {
        super.init(config);
        try { Class.forName("com.microsoft.sqlserver.jdbc.SQLServerDriver"); }
        catch(Exception e) { }
    }
    public String handleString(String s) {
        try{  byte bb[]=s.getBytes("iso-8859-1");
            s=new String(bb);
        }
        catch(Exception ee) { }
        return s;
    }
    public void doPost(HttpServletRequest request,HttpServletResponse response)
                                        throws ServletException,IOException {
        Connection con;
        PreparedStatement sql;
        Login loginBean=null;
        String backNews="";
        HttpSession session=request.getSession(true);
        try{  loginBean=(Login)session.getAttribute("login");
            if(loginBean==null) {
                loginBean=new Login();
```

```java
                        session.setAttribute("login",loginBean);
                    }
                }
                catch(Exception ee) {
                    loginBean=new Login();
                    session.setAttribute("login",loginBean);
                }
                String logname=request.getParameter("logname").trim(),
                password=request.getParameter("password").trim();
                boolean ok=loginBean.getSuccess();
                logname=handleString(logname);
                password=handleString(password);
                if(ok==true&&logname.equals(loginBean.getLogname())) {
                    backNews=logname+"已经登录了";
                    loginBean.setBackNews(backNews);
                }
                else {
                    String uri="jdbc:sqlserver://127.0.0.1:1433;DatabaseName=ComeHere";
                    boolean boo=(logname.length()>0)&&(password.length()>0);
                    try{    con=DriverManager.getConnection(uri,"sa","sa");
                        String condition="select * from member where logname =? and password =?";
                        sql=con.prepareStatement(condition);
                        if(boo) {
                            sql.setString(1,logname);
                            sql.setString(2,password);
                            ResultSet rs=sql.executeQuery();
                            boolean m=rs.next();
                            if(m==true) {
                                backNews="登录成功";
                                loginBean.setBackNews(backNews);
                                loginBean.setSuccess(true);
                                loginBean.setLogname(logname);
                            }
                            else {
                                backNews="您输入的用户名不存在，或密码不般配";
                                loginBean.setBackNews(backNews);
                                loginBean.setSuccess(false);
                                loginBean.setLogname(logname);
                                loginBean.setPassword(password);
                            }
                        }
                        else {
                            backNews="您输入的用户名不存在，或密码不般配";
                            loginBean.setBackNews(backNews);
                            loginBean.setSuccess(false);
                            loginBean.setLogname(logname);
                            loginBean.setPassword(password);
```

```
            }
            con.close();
        }
        catch(SQLException exp) {
            backNews=""+exp;
            loginBean.setBackNews(backNews);
            loginBean.setSuccess(false);
        }
    }
    RequestDispatcher dispatcher= request.getRequestDispatcher("/showLoginMess.jsp");
    dispatcher.forward(request, response);
}
public void doGet(HttpServletRequest request,HttpServletResponse response)
                    throws ServletException,IOException {
    doPost(request,response);
}
}
```

10.6 上传照片

用户可在该模块上传自己的照片。如果 member 中已经存有一幅照片，新上传的照片将替换原有的照片。用户在注册时，注册模块给会员的照片是默认图像 public.jpg。

该模块的模型 JavaBean 描述用户上传的照片文件的有关信息；视图部分由两个 JSP 页面构成，一个 JSP 页面负责提交图像文件到控制器，另一个 JSP 页面负责显示上传操作是否成功的信息；控制器负责将图像文件上传到服务器、将图像文件的名字写入数据库的 member 表中，必要时还需删除用户曾上传的图像文件。该 Servlet 还负责更新视图，使用户能看到上传操作的结果。另外，控制器能阻止未登录用户上传照片。

1. 模型（JavaBean）

下列 JavaBean 的实例用来描述上传文件的有关信息。

UploadFile.java
```
package mybean.data;
public class UploadFile {
    String fileName,savedFileName,
            backNews="";
    public void setFileName(String name) {
        fileName=name;
    }
    public String getFileName() {
        return fileName;
    }
    public void setSavedFileName(String name) {
        savedFileName=name;
    }
    public String getSavedFileName() {
```

```
            return savedFileName;
        }
        public String getBackNews() {
            return backNews;
        }
        public void setBackNews(String s) {
            backNews=s;
        }
    }
```

2．视图（JSP 页面）

本模块视图有两个 JSP 页面：upload.jsp 和 showUploadMess.jsp。upload.jsp 页面负责提供上传文件的表单；showUploadMess.jsp 负责显示上传文件的反馈信息。

upload.jsp（效果如图 10.7 所示）

图 10.7　选择上传的文件

```
<%@ page contentType="text/html;charset=GB2312" %>
<HTML><HEAD><%@ include file="head.txt" %></HEAD>
<BODY bgcolor=orange><CENTER>
  <P>文件将被上传到 D:\apache-tomcat-5.5.20\webapps\chapter10\image 中。
  <P>选择要上传的图像照片文件(名字不可以含有非 ASCII 码字符，比如汉字等)：
  <BR>
    <FORM action="helpUpload" method="post" ENCTYPE="multipart/form-data">
      <Input type=FILE name="fileName" size="40">
      <BR> <Input type="submit" name ="g" value="提交">
    </FORM></CENTER>
</BODY>
</HTML>
```

showUploadMess.jsp（效果如图 10.8 所示）

```
<%@ page contentType="text/html;charset=GB2312" %>
<%@ page import="mybean.data.UploadFile"%>
<jsp:useBean id="upFile" type="mybean.data.UploadFile" scope="request" />
<HTML><HEAD><%@ include file="head.txt" %></HEAD>
<HTML><BODY bgcolor=yellow >
<CENTER><FONT size=4 color=blue >
```

图 10.8 显示上传操作信息

```
<BR> <jsp:getProperty name="upFile" property="backNews" />
    </FONT>
<BR><FONT size=2>上传的文件名字：<jsp:getProperty name="upFile" property="fileName" />
    保存后的文件名字：<jsp:getProperty name="upFile" property="savedFileName" />
<BR> <IMG src=image/<jsp:getProperty name="upFile" property="savedFileName" />
        width=100 height=100>图像效果
    </IMG>
</FONT></CENTER>
</BODY></HTML>
```

3．控制器（Servlet）

该 Servlet 负责检查用户是否是登录用户，如果用户没有登录，控制器将把用户定向到登录页面 login.jsp；对于登录的用户，该 Servlet 负责把用户提交的图像文件保存到当前 Web 服务目录的特定子目录 image 中，即保存到目录 D:\apache-tomcat-5.5.20\webapps\chapter10\image 中，保存的图像文件名字是在用户上传的文件名字前面添加上用户的会员名。该 Servlet 同时负责将保存的图像文件名存入 member 表，然后将用户转发到 showUploadMess.jsp 页面查看上传操作的反馈信息。

HandleUpload.java

```java
package myservlet.control;
import mybean.data.*;
import java.sql.*;
import java.io.*;
import javax.servlet.*;
import javax.servlet.http.*;
public class HandleUpload extends HttpServlet {
    public void init(ServletConfig config) throws ServletException {
        super.init(config);
        try { Class.forName("com.microsoft.sqlserver.jdbc.SQLServerDriver"); }
        catch(Exception e){}
    }
    public void doPost(HttpServletRequest request,HttpServletResponse response)
                                            throws ServletException,IOException {
        HttpSession session=request.getSession(true);
        Login login=(Login)session.getAttribute("login");        //获取用户登录时的 Javabean
```

```java
            boolean ok=true;
            if(login==null) {
                ok=false;
                response.sendRedirect("login.jsp");          //重定向到登录页面
            }
            if(ok==true) {
                String logname=login.getLogname();
                uploadFileMethod(request,response,logname);  //接收上传文件
            }
        }
        public void uploadFileMethod(HttpServletRequest request,HttpServletResponse response,
                            String logname) throws ServletException,IOException {
            UploadFile upFile=new UploadFile();
            String backNews="";
            try{ HttpSession session=request.getSession(true);
                request.setAttribute("upFile",upFile);
                String tempFileName=(String)session.getId();
                File f1=new File(tempFileName);
                FileOutputStream o=new FileOutputStream(f1);
                InputStream in=request.getInputStream();
                byte b[]=new byte[10000];
                int n;
                while( (n=in.read(b))!=-1) {
                    o.write(b,0,n);
                }
                o.close();
                in.close();
                RandomAccessFile random=new RandomAccessFile(f1,"r");
                int second=1;              //读出 f1 的第 2 行，析取出上传文件的名字
                String secondLine=null;
                while(second<=2) {
                    secondLine=random.readLine();
                    second++;
                }
                int position=secondLine.lastIndexOf('\\'); //获取第 2 行中目录符号'\'最后出现的位置
                //截取文件名
                String fileName=secondLine.substring(position+1,secondLine.length()-1);
                byte   cc[]=fileName.getBytes("ISO-8859-1");
                fileName=new String(cc);
                fileName=fileName.replaceAll(" ","");
                //文件是否由字母或数字组成判断名字
                String checkedStr=fileName.substring(0,fileName.indexOf("."));
                boolean isLetterOrDigit=true;
                for(int i=0;i<checkedStr.length();i++) {
                    char c=checkedStr.charAt(i);
                    if(!((c<='z'&&c>='a')||(c<='Z'&&c>='A')||(c<='9'&&c>='0'))) {
                        isLetterOrDigit=false;
```

```
            break;
        }
    }
    if(isLetterOrDigit==false) {
        response.sendRedirect("upload.jsp");           //重定向到 upload.jsp 页面
    }
//保存文件名是上传文件名加会员名为前缀:
String savedFileName=logname.concat(fileName);
random.seek(0);
long    forthEndPosition=0;                            //获取第 4 行回车符号的位置
int forth=1;
while((n=random.readByte())!=-1&&(forth<=4)) {
    if(n=='\n') {
        forthEndPosition=random.getFilePointer();
        forth++;
    }
}
//根据客户上传文件的名字，将该文件存入磁盘
File dir=new File("D:/apache-tomcat-5.5.20/webapps/chaper10/image");
dir.mkdir();
//首先删除用户曾上传过的图像文件:
File file[]=dir.listFiles();
for(int k=0;k<file.length;k++) {
    if(file[k].getName().startsWith(logname))
        file[k].delete();
}
File savingFile= new File(dir,savedFileName);          //需要新保存的上传文件
RandomAccessFile random2=new RandomAccessFile(savingFile,"rw");
random.seek(random.length());
long endPosition=random.getFilePointer();
long mark=endPosition;
int j=1;
//确定出文件 f1 中包含客户上传的文件的内容的最后位置，即倒数第 6 行:
while((mark>=0)&&(j<=6)) {
    mark--;
    random.seek(mark);
    n=random.readByte();
    if(n=='\n') {
        endPosition=random.getFilePointer();
        j++;
    }
}
random.seek(forthEndPosition);
long startPoint=random.getFilePointer();
while(startPoint<endPosition-1) {
    n=random.readByte();
    random2.write(n);
```

```
                    startPoint=random.getFilePointer();
                }
                random2.close();
                random.close();
                String uri="jdbc:sqlserver://127.0.0.1:1433;DatabaseName=ComeHere";
                Connection con=DriverManager.getConnection(uri,"sa","sa");
                Statement sql=con.createStatement(ResultSet.TYPE_SCROLL_SENSITIVE,
                                            ResultSet.CONCUR_UPDATABLE);
                ResultSet rs=ql.executeQuery("SELECT * FROM members
                                            where logname = '"+logname+"'");
            if(rs.next()){
                if(isLetterOrDigit) {
                    rs.updateString(8,savedFileName);
                    int index=rs.getRow();
                    rs.absolute(index);
                    rs.updateRow();
                    backNews=fileName+"成功上传";
                    upFile.setFileName(fileName);
                    upFile.setSavedFileName(savedFileName);
                    upFile.setBackNews(backNews);
                }
            }
            con.close();
            f1.delete();
        }
        catch(Exception exp) {
            backNews=""+exp;
            upFile.setBackNews(backNews);
        }
        try{   RequestDispatcher
                        dispatcher= request.getRequestDispatcher("/showUploadMess.jsp");
            dispatcher.forward(request, response);
        }
        catch(Exception ee){ }
    }
    public void doGet(HttpServletRequest request,HttpServletResponse response)
                                            throws ServletException,IOException {
        doPost(request,response);
    }
}
```

10.7 浏览会员

该模块负责分页显示注册会员的信息,包括会员名、性别、照片等,同时提供查找功能,即用户可以查找某个会员的信息。该模块的模型 JavaBean 分别描述会员信息和分页信息;视图部分由三个 JSP 页面构成,一个 JSP 页面负责提交用户浏览会员信息的方式:分页浏览全部会员信息、浏览特定会员信息,另外两个 JSP 页面分别负责分页显示全体会员信息和显示特定会员信息;控制器 Servlet 使用 doPost()方法查询数据库表中的全部记录,并对记录进行分页处理,使用 doGet()方法查询数据库表中特定的记录。另外,控制器能阻止未登录用户浏览和查询会员信息。

1. 模型(JavaBean)

模型有两个 JavaBean。

(1)会员信息

下列 JavaBean 的实例用来描述会员信息。

MemberInform.java

```java
package mybean.data;
public class MemberInform {
    String logname="",sex="",email="",
           phone="",address="",
           message="",pic,backNews="";
    int age;
    public void setLogname(String name) {
        logname=name;
    }
    public String getLogname(){
        return logname;
    }
    public void setAge(int n) {
        age=n;
    }
    public int getAge(){
        return age;
    }
    public void setSex(String s) {
        sex=s;
    }
    public String getSex(){
      return sex;
    }
    public void setEmail(String em) {
        email=em;
    }
    public String getEmail(){
        return email;
    }
    public void setPhone(String ph) {
        phone=ph;
    }
    public String getPhone(){
        return phone;
    }
```

```java
        public String getMessage(){
            return message;
        }
        public void setMessage(String m) {
            message=m;
        }
        public String getPic(){
            return pic;
        }
        public void setPic(String s) {
            pic=s;
        }
        public String getBackNews(){
            return backNews;
        }
        public void setBackNews(String s) {
            backNews=s;
        }
    }
```

（2）分页信息

下列 JavaBean 的实例用来描述记录分页信息。

MemberInform.java

```java
        package mybean.data;
        import com.sun.rowset.*;
        public class ShowByPage {
            CachedRowSetImpl rowSet=null;       //存储表中全部记录的行集对象
            int pageSize=10;                    //每页显示的记录数
            int pageAllCount=0;                 //分页后的总页数
            int showPage=1;                     //当前显示页
            StringBuffer presentPageResult;     //显示当前页内容
            public void setRowSet(CachedRowSetImpl set) {
                rowSet=set;
            }
            public CachedRowSetImpl getRowSet(){
                return rowSet;
            }
            public void setPageSize(int size) {
                pageSize=size;
            }
            public int getPageSize(){
                return pageSize;
            }
            public int getPageAllCount(){
                return pageAllCount;
            }
            public void setPageAllCount(int n) {
```

```
                pageAllCount=n;
            }
            public void setShowPage(int n) {
                showPage=n;
            }
            public int getShowPage(){
                return showPage;
            }
            public void setPresentPageResult(StringBuffer p) {
                presentPageResult=p;
            }
            public StringBuffer getPresentPageResult() {
                return presentPageResult;
            }
        }
```

2. 视图（JSP 页面）

本模块视图有三个 JSP 页面：choiceLookType.jsp、showLookedMember.jsp 和 showAllMember.jsp。choiceLookType.jsp 负责将浏览会员的方式提交给控制器，showLookedMember.jsp 负责显示被查找的会员的信息，showAllMember.jsp 负责分页显示全体会员的信息。

choiceLookType.jsp（效果如图 10.9 所示）

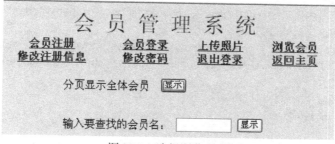

图 10.9　选择浏览方式

```
<%@ page contentType="text/html;charset=GB2312" %>
<HTML><HEAD><%@ include file="head.txt" %></HEAD>
<BODY bgcolor=cyan><CENTER><FONT size=3>
    <TABLE>
      <FORM action="helpShowMember" method="post" name="form">
        <BR>分页显示全体会员
        <Input type="hidden" value="1" name="showPage" size=6>
        <Input type="submit" value="显示" name="submit">
    </FORM>
    <FORM action="helpShowMember" method="get" name="form">
        <BR>输入要查找的会员名：
        <Input type="text"    name="logname" size=6>
        <Input type="submit" value="显示" name="submit">
    </FORM></CENTER>
</BODY></HTML>
```

showAllMember.jsp（效果如图 10.10 所示）

图 10.10 分页显示会员信息

```
<%@ page contentType="text/html;charset=GB2312" %>
<%@ page import="mybean.data.ShowByPage" %>
<jsp:useBean id="show" type="mybean.data.ShowByPage" scope="session"/>
<HTML><%@ include file="head.txt" %></HEAD>
<BODY ><CENTER>
<P>显示会员信息
  <BR>每页最多显示<jsp:getProperty  name= "show"  property="pageSize"  />条信息
   <BR>当前显示第<FONT color=blue>
       <jsp:getProperty  name= "show"  property="showPage"  /></FONT>页,共有
        <FONT color=blue><jsp:getProperty  name= "show"  property="pageAllCount"  />
         </FONT>页。
  <BR>当前显示的内容是:
<TABLE border=2>
    <TR>
      <TH>会员名</TH><TH>性别</TH><TH>年龄</TH><TH>电话</TH>
          <TH>email</TH><TH>简历</TH><TH>照片</TH>
    </TR>
<jsp:getProperty  name= "show"  property="presentPageResult" />
</TABLE>
<BR>单击"前一页"或"下一页"按钮查看信息
<TABLE>
<TR><TD><FORM action="helpShowMember" method=post>
        <Input type=hidden name="showPage" value="<%=show.getShowPage()-1 %>" >
         <Input type=submit name="g" value="前一页">
        </FORM>
</TD>
    <TD><FORM action="helpShowMember" method=post>
         <Input type=hidden name="showPage" value="<%=show.getShowPage()+1 %>" >
         <Input type=submit name="g" value="后一页">
        </FORM>
</TD>
```

```
        <TD> <FORM action="helpShowMember" method=post>
            输入页码：<Input type=text name="showPage" size=5 >
            <Input type=submit name="g" value="提交">
            </FORM>
        </TD>
      </TR>
    </TABLE>
  </CENTEr>
</BODY></HTML>
```

showLookedMember.jsp（效果如图 10.11 所示）

图 10.11 显示某个会员的信息

```
<%@ page contentType="text/html;charset=GB2312" %>
<%@ page import="mybean.data.MemberInform" %>
<jsp:useBean id="inform" type="mybean.data.MemberInform" scope="request" />
<CENTER>
<HTML><BODY bgcolor=pink><%@ include file="head.txt" %></HEAD>
<FONT size=3>
<TABLE border=2>
  <TR>
    <TH>会员名</TH><TH>性别</TH><TH>年龄</TH><TH>电话</TH>
    <TH>email</TH><TH>简历</TH><TH>照片</TH>
  </TR>
  <TR>
    <TD><jsp:getProperty  name= "inform"   property="logname" /></TD>
    <TD><jsp:getProperty  name= "inform"   property="sex" /></TD>
    <TD><jsp:getProperty  name= "inform"   property="age" /></TD>
    <TD><jsp:getProperty  name= "inform"   property="phone" /></TD>
    <TD><jsp:getProperty  name= "inform"   property="email" /></TD>
    <TD><jsp:getProperty  name= "inform"   property="message" /></TD>
    <TD><IMG src=image/<jsp:getProperty   name= "inform" property="pic" />></IMG></TD>
</TABLE>
</CENTER>
</BODY></HTML>
```

3. 控制器（Servlet）

Servlet 使用 doPost()方法查询数据库表中的全部记录，对记录进行分页处理，使用 doGet() 方法查询数据库表中特定的记录。另外，控制器能阻止未登录用户浏览和查询会员信息。

HandleDatabase.java

```java
package myservlet.control;
import mybean.data.*;
import com.sun.rowset.*;
import java.sql.*;
import java.io.*;
import javax.servlet.*;
import javax.servlet.http.*;
public class HandleDatabase extends HttpServlet {
    CachedRowSetImpl rowSet=null;
    public void init(ServletConfig config) throws ServletException {
        super.init(config);
        try { Class.forName("com.microsoft.sqlserver.jdbc.SQLServerDriver"); }
        catch(Exception e) { }
    }
    public void doPost(HttpServletRequest request,HttpServletResponse response)
                                           throws ServletException,IOException {
        HttpSession session=request.getSession(true);
        Login login=(Login)session.getAttribute("login");      //获取用户登录时的 Javabean
        boolean ok=true;
        if(login==null) {
           ok=false;
           response.sendRedirect("login.jsp");                 //重定向到登录页面
        }
        if(ok==true) {
            continueDoPost(request,response);
        }
    }
    public void continueDoPost(HttpServletRequest request,HttpServletResponse response)
                                            throws ServletException,IOException {
        HttpSession session=request.getSession(true);
        Connection con=null;
        StringBuffer presentPageResult=new StringBuffer();
        ShowByPage showBean=null;
        try{   showBean=(ShowByPage)session.getAttribute("show");
              if(showBean==null) {
                  showBean=new ShowByPage();     //创建 Javabean 对象
                  session.setAttribute("show",showBean);
              }
        }
        catch(Exception exp) {
             showBean=new ShowByPage();
             session.setAttribute("show",showBean);
        }
        showBean.setPageSize(3);                    //每页显示 3 条记录
        int showPage=Integer.parseInt(request.getParameter("showPage"));
        if(showPage>showBean.getPageAllCount())
             showPage=1;
```

```java
            if(showPage<=0)
                showPage=showBean.getPageAllCount();
            showBean.setShowPage(showPage);
            int pageSize=showBean.getPageSize();
            String uri="jdbc:sqlserver://127.0.0.1:1433;DatabaseName= ComeHere";
            try{  con=DriverManager.getConnection(uri,"sa","sa");
                Statement sql=con.createStatement(ResultSet.TYPE_SCROLL_SENSITIVE,
                                            ResultSet.CONCUR_READ_ONLY);
                ResultSet rs=sql.executeQuery("SELECT * FROM member");
                rowSet=new CachedRowSetImpl();        //创建行集对象
                rowSet.populate(rs);
                con.close();                          //关闭连接
                showBean.setRowSet(rowSet);           //数据存储在 showBean 中
                rowSet.last();
                int m=rowSet.getRow();                //总行数
                int n=pageSize;
                int pageAllCount=((m%n)==0)?(m/n):(m/n+1);
                showBean.setPageAllCount(pageAllCount);//数据存储在 showBean 中
                presentPageResult=show(showPage,pageSize,rowSet);
                showBean.setPresentPageResult(presentPageResult);
            }
            catch(SQLException exp){ }
            RequestDispatcher dispatcher= request.getRequestDispatcher("/showAllMember.jsp");
            dispatcher.forward(request, response);
    }
    public StringBuffer show(int page,int pageSize,CachedRowSetImpl rowSet) {
        StringBuffer str=new StringBuffer();
        try{   rowSet.absolute((page-1)*pageSize+1);
            for(int i=1;i<=pageSize;i++) {
                str.append("<tr>");
                str.append("<td>"+rowSet.getString(1)+"</td>");
                str.append("<td>"+rowSet.getString(3)+"</td>");
                str.append("<td>"+rowSet.getString(4)+"</td>");
                str.append("<td>"+rowSet.getString(5)+"</td>");
                str.append("<td>"+rowSet.getString(6)+"</td>");
                str.append("<td>"+rowSet.getString(7)+"</td>");
                String s="<img src=image/"+rowSet.getString(8)+" width=100 height=100/>";
                str.append("<td>"+s+"</td>");
                str.append("</tr>");
                rowSet.next();
            }
        }
        catch(SQLException exp) { }
        return str;
    }
    public void doGet(HttpServletRequest request,HttpServletResponse response)
                                    throws ServletException,IOException {
```

```
            HttpSession session=request.getSession(true);
            Login login=(Login)session.getAttribute("login");    //获取用户登录时的Javabean
            boolean ok=true;
            if(login==null) {
               ok=false;
               response.sendRedirect("login.jsp");              //重定向到登录页面
            }
            if(ok==true) {
              continueDoGet(request,response);
            }
         }
         public void continueDoGet(HttpServletRequest request,HttpServletResponse response)
                                                throws ServletException,IOException {
            MemberInform inform=new MemberInform();
            request.setAttribute("inform",inform);
            String logname=request.getParameter("logname");
            Connection con=null;
            String uri="jdbc:sqlserver://127.0.0.1:1433;DatabaseName= ComeHere";
            try{  con=DriverManager.getConnection(uri,"sa","sa");
                  Statement sql=con.createStatement();
                  ResultSet rs=sql.executeQuery("SELECT * FROM member
                                           WHERE logname = '"+logname+"'");
                  if(rs.next()) {
                      inform.setLogname(rs.getString(1));
                      inform.setSex(rs.getString(3));
                      inform.setAge(rs.getInt(4));
                      inform.setPhone(rs.getString(5));
                      inform.setEmail(rs.getString(6));
                      inform.setMessage(rs.getString(7));
                      inform.setPic(rs.getString(8));
                      inform.setBackNews("查询到的会员信息：");
                  }
                  con.close();
                  RequestDispatcher dispatcher=
                                  request.getRequestDispatcher("/showLookedMember.jsp");
                  dispatcher.forward(request, response);
            }
            catch(SQLException exp) {
                inform.setBackNews(""+exp);System.out.println("ok1"+exp);
            }
         }
      }
```

10.8 修改密码

登录的用户可在该模块修改密码。该模块的模型 JavaBean 描述密码的有关信息；该模块视图部分由两个 JSP 页面构成，一个 JSP 页面负责提交用户的新旧密码到控制器，另一个 JSP

页面负责显示修改是否成功的信息。该模块的控制器 Servlet 负责修改密码。

1. 模型（JavaBean）

下列 JavaBean 的实例用来描述修改密码有关信息。

Password.java
```java
package mybean.data;
public class Password {
    String oldPassword,newPassword,backNews="";
    public void setNewPassword(String pw) {
        newPassword=pw;
    }
    public String getnewPassword(){
        return newPassword;
    }
    public void setOldPassword(String pw) {
        oldPassword=pw;
    }
    public String getOldPassword() {
        return oldPassword;
    }
    public String getBackNews() {
        return backNews;
    }
    public void setBackNews(String s) {
        backNews=s;
    }
}
```

2. 视图（JSP 页面）

本模块视图有 2 个 JSP 页面：modifyPassword.jsp、showNewPasswor.jsp。modifyPassword.jsp 页面负责提供输入密码界面，showNewPasswor.jsp 负责显示修改密码的反馈信息。

modifyPassword.jsp（效果如图 10.12 所示）
```jsp
<%@ page contentType="text/html;charset=GB2312" %>
<HTML><HEAD><%@ include file="head.txt" %></HEAD>
<BODY bgcolor=pink><Font size=2><CENTER>
<P>输入您的密码：
<FORM action="helpModifyPassword" Method="post">
<BR>当前密码:<Input type=text name="oldPassword">
<BR>新密码: <Input type=password name="newPassword">
<BR><Input type=submit name="g" value="提交">
</FORM></CENTER>
</BODY></HTML>
```

showNewPasswor.jsp（效果如图 10.13 所示）
```jsp
<%@ page contentType="text/html;charset=GB2312" %>
<%@ page import="mybean.data.Password"%>
<jsp:useBean id="password" type="mybean.data.Password" scope="request" />
```

图 10.12 修改密码   图 10.13 显示修改密码结果

```
<HTML><HEAD><%@ include file="head.txt" %></HEAD>
<HTML><BODY bgcolor=yellow >
<CENTER>
   <jsp:getProperty name="password" property="backNews" />
   <BR>您的新密码: <jsp:getProperty name="password" property="newPassword" />
   <BR>您的旧密码: <jsp:getProperty name="password" property="oldPassword" />
</FONT></CENTER>
</BODY></HTML>
```

3．控制器（Servlet）

控制器负责连接数据库，根据当前用户注册的会员名修改 member 表中该会员的 password 字段的值，并转发修改信息到 showModifyMess.jsp 页面。另外，控制器能阻止未登录用户浏览进行修改密码操作。

HandlePassword.java

```java
package myservlet.control;
import mybean.data.*;
import java.sql.*;
import java.io.*;
import javax.servlet.*;
import javax.servlet.http.*;
public class HandlePassword extends HttpServlet {
    public void init(ServletConfig config) throws ServletException {
        super.init(config);
        try { Class.forName("com.microsoft.sqlserver.jdbc.SQLServerDriver"); }
        catch(Exception e){ }
    }
    public void doPost(HttpServletRequest request,HttpServletResponse response)
                                            throws ServletException,IOException {
        HttpSession session=request.getSession(true);
        Login login=(Login)session.getAttribute("login");       //获取用户登录时的 Javabean
        boolean ok=true;
        if(login==null) {
           ok=false;
           response.sendRedirect("login.jsp");                  //重定向到登录页面
        }
        if(ok==true) {
           continueWork(request,response);
```

```java
        }
    }
    public void continueWork(HttpServletRequest request,HttpServletResponse response)
                                        throws ServletException,IOException {
        HttpSession session=request.getSession(true);
        Login login=(Login)session.getAttribute("login");
        Connection con=null;
        String logname=login.getLogname();
        Password passwordBean=new Password();
        request.setAttribute("password",passwordBean);
        String oldPassword=request.getParameter("oldPassword");
        String newPassword=request.getParameter("newPassword");
        String uri="jdbc:sqlserver://127.0.0.1:1433;DatabaseName= ComeHere";
        try{  con=DriverManager.getConnection(uri,"sa","sa");
            Statement sql=con.createStatement();
            ResultSet rs= sql.executeQuery("SELECT * FROM member
                        where logname='"+logname+"'And password='"+oldPassword+"'");
            if(rs.next()) {
                String updateString="UPDATE member SET password='"+
                                newPassword+"' where logname='"+logname+"'";
                int m=sql.executeUpdate(updateString);
                if(m==1) {
                    passwordBean.setBackNews("密码更新成功");
                    passwordBean.setOldPassword(oldPassword);
                    passwordBean.setNewPassword(newPassword);
                }
                else {
                    passwordBean.setBackNews("密码更新失败");
                }
            }
            else {
                passwordBean.setBackNews("密码更新失败");
            }
        }
        catch(SQLException exp) {
            passwordBean.setBackNews("密码更新失败"+exp);
        }
        RequestDispatcher dispatcher=
                            request.getRequestDispatcher("/showNewPassword.jsp");
        dispatcher.forward(request, response);
    }
    public void doGet(HttpServletRequest request,HttpServletResponse response)
                                        throws ServletException,IOException {
        doPost(request,response);
    }
}
```

10.9 修改注册信息

用户可在该模块修改曾注册的个人信息。该模块的模型 JavaBean 描述用户修改的信息；该模块视图部分由三个 JSP 页面构成，第一个 JSP 页面负责确认是否进行修改，第二个 JSP 页面负责提交用户的修改信息到控制器，最后一个 JSP 页面负责显示修改是否成功的信息。该模块的控制器有两个 Servlet，一个负责读取用户曾注册的信息；另一个负责修改曾注册的信息，两个 Servlet 都能阻止未登录用户使用该模块。

1. 模型（JavaBean）

模型中需要两个 JavaBean，一个是在 10.4 节的模型中的 Register.java，另一个是下列 ModifyMessage.java，该类的实例用来描述用户的修改信息。

ModifyMessage.java

```java
package mybean.data;
public class ModifyMessage {
    String logname="",newSex="",
        newEmail="",newPhone="",newMessage="",backNews;
    int newAge;
    public void setLogname(String name) {
        logname=name;
    }
    public String getLogname() {
        return logname;
    }
    public void setNewAge(int n) {
        newAge=n;
    }
    public int getNewAge(){
        return newAge;
    }
    public void setNewSex(String s) {
        newSex=s;
    }
    public String getNewSex() {
        return newSex;
    }
    public void setNewEmail(String em) {
        newEmail=em;
    }
    public String getNewEmail() {
        return newEmail;
    }
    public void setNewPhone(String ph) {
        newPhone=ph;
    }
    public String getNewPhone(){
```

```
            return newPhone;
        }
          public String getNewMessage(){
            return newMessage;
        }
        public void setNewMessage(String m) {
            newMessage=m;
        }
        public String getBackNews(){
            return backNews;
        }
        public void setBackNews(String s) {
            backNews=s;
        }
    }
```

2．视图（JSP 页面）

本模块视图有 san 个 JSP 页面：choiceModifyMess.jsp，inputNewMess.jsp 和 showModifyMess.jsp。用户选择修改信息视图后，首先进入 choiceModifyMess.jsp 页面，从中确认是否开始修改操作；inputNewMess.jsp 页面负责提供修改信息界面，showModifyMess.jsp 负责显示修改反馈信息。当用户访问 choiceModifyMess 页面时，如果确认要修改自己的注册信息，将先被定向到控制器中的一个 Servlet，如果是登录的用户，该 Servlet 将查询该用户曾注册的信息，并将该登录用户重新定向到 inputNewMess.jsp 页面；如果用户没有登录，将被定向到登录页面 login.jsp。

choiceModifyMess.jsp（效果如图 10.14 所示）

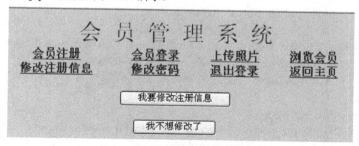

图 10.14　选择修改或放弃修改

```
<%@ page contentType="text/html;charset=GB2312" %>
<CENTER>
<HTML><BODY bgcolor=pink><%@ include file="head.txt" %></HEAD>
 <FORM action="helpGetOldMess" name=form>
   <Input type=submit value="我要修改注册信息">
 </FORM>
 <FORM action="index.jsp" name=form>
   <Input type=submit value="我不想修改了">
 </FORM>
</CENTER>
</BODY></HTML>
```

inputModifyMess.jsp（效果如图 10.15 所示）

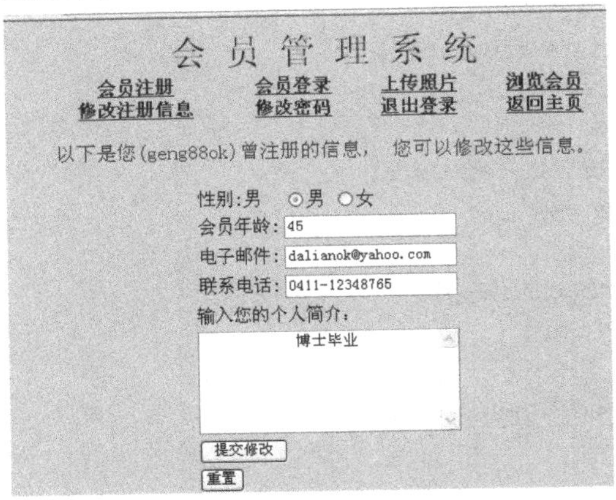

图 10.15　修改信息

```
<%@ page contentType="text/html;charset=GB2312" %>
<HTML><HEAD><%@ include file="head.txt" %></HEAD>
<%@ page import="mybean.data.Register"%>
<jsp:useBean id="register" type="mybean.data.Register" scope="request" />
<HTML><BODY bgcolor=pink><CENTER>
<P>
<FONT color=blue size=4>
    以下是您(<jsp:getProperty name="register" property="logname"/>)曾注册的信息，
    您可以修改这些信息。
</FONT>
<FONT size=2>
<FORM action="helpModifyMess" name=form>
<TABLE>
   <TR><TD>性别:<jsp:getProperty name="register"  property="sex"  /></TD>
      <TD><Input type=radio  name="newSex" checked="o" value="男">男
         <Input type=radio  name="newSex" value="女">女
     </TD>
   </TR>
   <TR><TD>会员年龄:</TD><TD><Input type=text name="newAge"
         value=<jsp:getProperty name="register"  property="age"  />></TD></TR>
   <TR><TD>电子邮件:</TD><TD><Input type=text name="newEmail"
         value=<jsp:getProperty name="register"  property="email"  />></TD></TR>
   <TR><TD>联系电话:</TD><TD><Input type=text name="newPhone"
         value=<jsp:getProperty name="register"  property="phone"  />></TD></TR>
</TABLE>
<TABLE>
   <TR><TD>输入您的个人简介：</TD></TR>
   <TR>
      <TD><TextArea name="newMessage" Rows="6" Cols="30">
           <jsp:getProperty name="register" property="message"/>
         </TextArea></TD>
   </TR>
```

```
            <TR><TD><Input type=submit name="g" value="提交修改"></TD></TR>
            <TR><TD><Input type=reset value="重置"></TD> </TR>
        </TABLE><FONT></CENTER>
    </BODY></HTML>
```

showModifyMess.jsp（效果如图 10.16 所示）

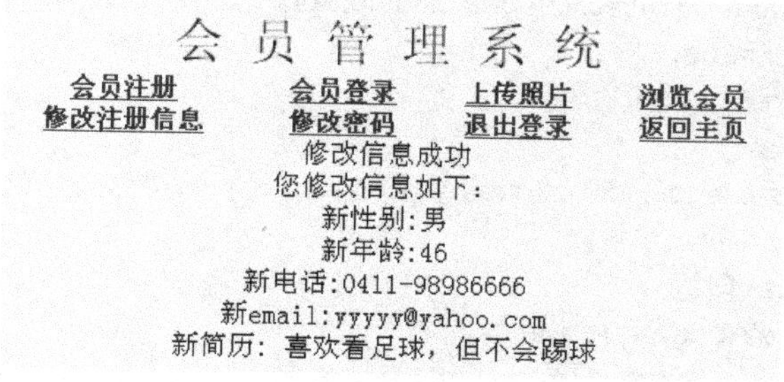

图 10.16 显示修改结果

```
<%@ page contentType="text/html;charset=GB2312" %>
<%@ page import="mybean.data.ModifyMessage"%>
<jsp:useBean id="modify" type="mybean.data.ModifyMessage" scope="request" />
<HTML><HEAD><%@ include file="head.txt" %></HEAD>
<HTML><BODY bgcolor=yellow ><FONT size=4>
<CENTER>
    <jsp:getProperty name="modify" property="backNews" />
    <BR>您修改信息如下：
    <BR>新性别:<jsp:getProperty name="modify" property="newSex" />
    <BR>新年龄:<jsp:getProperty name="modify" property="newAge" />
    <BR>新电话:<jsp:getProperty name="modify" property="newPhone" />
    <BR>新 email:<jsp:getProperty name="modify" property="newEmail" />
    <BR>新简历:<jsp:getProperty name="modify" property="newMessage" />
</FONT></CENTER>
</BODY></HTML>
```

3. 控制器（Servlet）

（1）查询注册信息的 Servlet

该 Servlet 负责连接数据库，查询 member 表，查询当前登录会员的注册信息（不包括密码和照片），然后将用户转发到 inputModifyMess.jsp 页面。

GetOldMess.java

```java
package myservlet.control;
import mybean.data.*;
import java.sql.*;
import java.io.*;
import javax.servlet.*;
import javax.servlet.http.*;
public class GetOldMess extends HttpServlet {
    public void init(ServletConfig config) throws ServletException {
        super.init(config);
        try{ Class.forName("com.microsoft.sqlserver.jdbc.SQLServerDriver"); }
```

```java
          catch(Exception e) { }
    }
    public void doPost(HttpServletRequest request,HttpServletResponse response)
                                            throws ServletException,IOException {
          HttpSession session=request.getSession(true);
          Login login=(Login)session.getAttribute("login");   //获取用户登录时的Javabean
          boolean ok=true;
          if(login==null) {
             ok=false;
             response.sendRedirect("login.jsp");              //重定向到登录页面
          }
          if(ok==true) {
             continueWork(request,response);
          }
    }
    public void continueWork(HttpServletRequest request,HttpServletResponse response)
                                            throws ServletException,IOException {
          HttpSession session=request.getSession(true);
          Login login=(Login)session.getAttribute("login");
          Connection con=null;
          String logname=login.getLogname();
          Register register=new Register();
          request.setAttribute("register",register);
          String uri="jdbc:sqlserver://127.0.0.1:1433;DatabaseName= ComeHere";
          try{  con=DriverManager.getConnection(uri,"sa","sa");
             Statement sql=con.createStatement();
             ResultSet rs=
             sql.executeQuery("SELECT * FROM member where logname='"+logname+"'");
             if(rs.next()){
                register.setLogname(rs.getString(1));
                register.setSex(rs.getString(3));
                register.setAge(rs.getInt(4));
                register.setPhone(rs.getString(5));
                register.setEmail(rs.getString(6));
                register.setMessage(rs.getString(7));
                register.setBackNews("您原来的注册信息:");
             }
          }
          catch(SQLException exp) { register.setBackNews(""+exp); }
          RequestDispatcher dispatcher= request.getRequestDispatcher("/inputModifyMess.jsp");
          dispatcher.forward(request, response);
    }
      public void doGet(HttpServletRequest request,HttpServletResponse response)
                                            throws ServletException,IOException {
          doPost(request,response);
      }
}
```

(2) 负责修改注册信息的 Servlet

该 Servlet 负责连接数据库，将用户提交的新的信息写入到 member 表，并将用户转发到 showModifyMess.jsp 页面查看修改反馈信息。

HandleModifyMess.java

```java
package myservlet.control;
import mybean.data.*;
import java.sql.*;
import java.io.*;
import javax.servlet.*;
import javax.servlet.http.*;
public class HandleModifyMess extends HttpServlet {
    public void init(ServletConfig config) throws ServletException {
        super.init(config);
        try { Class.forName("com.microsoft.sqlserver.jdbc.SQLServerDriver"); }
        catch(Exception e) { }
    }
    public String handleString(String s) {
        try{   byte bb[]=s.getBytes("iso-8859-1");
            s=new String(bb);
        }
        catch(Exception ee) { }
        return s;
    }
    public void doPost(HttpServletRequest request,HttpServletResponse response)
                                            throws ServletException,IOException {
        HttpSession session=request.getSession(true);
        Login login=(Login)session.getAttribute("login");        //获取用户登录时的Javabean
        boolean ok=true;
        if(login==null) {
           ok=false;
           response.sendRedirect("login.jsp");                   //重定向到登录页面
        }
        if(ok==true) {
           continueDoPost(request,response);
        }
    }
    public void continueDoPost(HttpServletRequest request,HttpServletResponse response)
                                            throws ServletException,IOException {
        HttpSession session=request.getSession(true);
        Login login=(Login)session.getAttribute("login");
        String logname=login.getLogname();
        Connection con;
        PreparedStatement sql;
        ModifyMessage modify=new ModifyMessage();
        request.setAttribute("modify",modify);
        String sex=request.getParameter("newSex").trim(),
```

```java
            email=request.getParameter("newEmail").trim(),
            phone=request.getParameter("newPhone").trim(),
            message=request.getParameter("newMessage");
            int age=Integer.parseInt(request.getParameter("newAge").trim());
            String uri="jdbc:sqlserver://127.0.0.1:1433;DatabaseName=ComeHere";
            String backNews="";
            try{ con=DriverManager.getConnection(uri,"sa","sa");
                String updateCondition=
                "UPDATE member SET sex=?,age=?,phone=?,email=?,message=?
                                                            WHERE logname=?";
                sql=con.prepareStatement(updateCondition);
                sql.setString(1,handleString(sex));
                sql.setInt(2,age);
                sql.setString(3,phone);
                sql.setString(4,handleString(email));
                sql.setString(5,handleString(message));
                sql.setString(6,logname);
                int m=sql.executeUpdate();
                if(m==1) {
                    backNews="修改信息成功";
                    modify.setBackNews(backNews);
                    modify.setLogname(logname);
                    modify.setNewAge(age);
                    modify.setNewSex(handleString(sex));
                    modify.setNewEmail(handleString(email));
                    modify.setNewPhone(phone);
                    modify.setNewMessage(handleString(message));
                }
                else {
                    backNews="信息填写不完整或信息中有非法字符";
                    modify.setBackNews(backNews);
                }
                con.close();
            }
            catch(SQLException exp) { modify.setBackNews(""+exp);  }
            RequestDispatcher dispatcher= request.getRequestDispatcher("/showModifyMess.jsp");
            dispatcher.forward(request, response);
        }
        public void doGet(HttpServletRequest request,HttpServletResponse response)
                                                    throws ServletException,IOException {
            doPost(request,response);
        }
    }
```

10.10 退出登录

该模块只有一个控制器，其中的 Servlet 负责销毁用户的 session 对象，导致登录失效。

HandleExit.java

```java
package myservlet.control;
import mybean.data.*;
import java.sql.*;
import java.io.*;
import javax.servlet.*;
import javax.servlet.http.*;
public class HandleExit extends HttpServlet {
    public void init(ServletConfig config) throws ServletException {
        super.init(config);
        try {
            Class.forName("com.microsoft.sqlserver.jdbc.SQLServerDriver");
        }
        catch(Exception e) { }
    }
    public String handleString(String s) {
        try{
            byte bb[]=s.getBytes("iso-8859-1");
            s=new String(bb);
        }
        catch(Exception ee) { }
      return s;
    }
    public void doPost(HttpServletRequest request,HttpServletResponse response)
                                            throws ServletException,IOException {
        HttpSession session=request.getSession(true);
        Login login=(Login)session.getAttribute("login");      //获取用户登录时的 JavaBean
        boolean ok=true;
        if(login==null) {
            ok=false;
            response.sendRedirect("login.jsp");             //重定向到登录页面
        }
        if(ok==true) {
            continueDoPost(request,response);
        }
    }
    public void continueDoPost(HttpServletRequest request,HttpServletResponse response)
                                            throws ServletException,IOException {
        HttpSession session=request.getSession(true);
        session.invalidate();                            //销毁用户的 session 对象
        response.sendRedirect("index.jsp");              //返回主页
    }
    public void doGet(HttpServletRequest request,HttpServletResponse response)
                                            throws ServletException,IOException {
        doPost(request,response);
    }
}
```

第 11 章

实训二：网上书城

本章导读

- 知识点：掌握 Web 应用中常见基本模块的开发方法。
- 重点：掌握 JSP+Tag 模式在设计 Web 应用中的重要性。
- 难点：掌握"确认订单"模块的设计。
- 关键实践：对"网上书城"进行调试。

第 11 章设计实现一个网上图书查询与订购系统，其目的是掌握一般 Web 应用中常用基本模块的开发方法。系统采用 JSP+Tag 模式实现各模块，重点突出 JSP 页面和 Tag 文件的作用（建议读者简单复习 3.7 节有关 Tag 文件的内容），尽可能让页面简洁。学生在阅读本章给出的系统基本模块后，不仅可以进行页面的美化工作，而且也可以增加更多的功能模块，以便熟悉 JSP+Tag 的开发模式。

JSP+Tag 模式与 MVC 模式都是 Web 开发中的常用模式，应根据项目的大小和复杂程度选择其一，一般来说，JSP+Tag 模式适合结构相对简单的 Web 项目，而 MVC 模式适合结构相对复杂的 Web 项目。读者在熟悉掌握了本章的各个模块后，可以参考第 10 章，将本章的开发模式更改为 JSP+JavaBean+Servlet 模式，即 MVC 模式，以便熟练使用两种模式开发 Web 项目。本章使用的 JSP 引擎为 Tomcat6.0，数据库管理系统是 Microsoft Access 2003，数据库连接操作使用 JDBC-ODBC 桥接器方式。

11.1 系统主要模块

系统主要模块如图 11.1 所示。

① 注册：用户填写表单，包括注册的用户名、E-mail 地址等信息。如果输入的用户名已经被其他用户注册使用，系统提示新用户更改自己的注册用户名。

② 登录：输入注册的用户名、密码。如果用户输入的用户名或密码有错误，系统将显示错误信息。

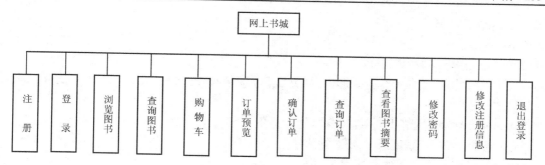

图 11.1 系统主要模块

③ 浏览图书：用户不必注册并登录可直接进入该页面浏览图书的相关信息，但是当用户将欲购买的图书放入自己的购物车时，系统将要求用户必须登录。

④ 查询图书：用户不必注册并登录可直接进入该页面，在该页面用户可以按图书的 ISBN、名称或作者进行查询操作，但是当用户将欲购买的图书放入自己的购物车时，系统将要求用户必须登录。

⑤ 购物车：登录的用户可以将欲购买的图书放入购物车。

⑥ 订单预览：登录的用户可以预览自己的订单。

⑦ 确认订单：登录的用户可以确认自己的订单。

⑧ 查询订单：登录的用户可以查询自己的全部订单。

⑨ 查看图书摘要：用户（不必登录），可以查看图书的内容摘要。

⑩ 修改密码：成功登录的会员可以在该页面修改自己的登录密码，如果用户直接进入该页面或没有成功登录就进入该页面，将被链接到"会员登录"页面。

⑪ 修改注册信息：成功登录的会员可以在该页面修改自己的注册信息，比如联系电话、通信地址等，如果用户直接进入该页面或没有成功登录就进入该页面，将被链接到"会员登录"页面。

⑫ 退出登录：成功登录的用户可以使用该模块退出登录。

用户与各个模块的主要关系如图 11.2 所示。

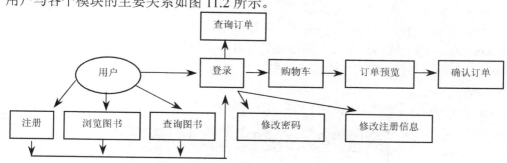

图 11.2 用户与主要模块的关系

11.2 数据库设计与连接

1. 数据库设计

使用 MicroSoft Access 2003 建立一个数据库 bookshop，该库共有三个表，以下是这些表的名称、结构和用途。

（1）user 表

表名：user。结构：如图 11.3 所示。

用途：存储用户的注册信息。即会员的注册信息存入 usr 表中，usr 表的主键是 logname，各字段值的说明如下：

logname：存储注册的用户名（属性是字符型）。

password：存储登录密码（属性是字符型）。

phone：存储电话（属性是字符型）。

email：电子邮件（属性是字符型）。

address：存储通信地址（属性是字符型）。

realname：存储真实姓名（属性是字符型）。

（2）bookForm 表

表名：bookForm。结构：如图 11.4 所示。

user ：表

	字段名称	数据类型
🔑	logname	文本
	password	文本
	phone	文本
	email	文本
	address	文本
	realname	文本

图 11.3　user 表

bookForm ：表

	字段名称	数据类型
	bookPic	文本
🔑	bookISBN	文本
	bookName	文本
	bookAuthor	文本
	bookPrice	数字
	bookPublish	文本
	bookAbstract	文本

图 11.4　bookForm 表

用途：存储图书信息，bookForm 表的主键是 bookISBN，各字段值的说明如下：

bookPic：存储与图书相关的一幅图像文件的名字（属性是字符型）。

bookISBN：图书的 ISBN（属性是字符型）。

bookName：图书的名称（属性是字符型）。

bookAuthor：图书的作者（属性是字符型）。

bookPrice：图书的的价格（属性是单精度浮点型）。

bookPublish：图书的的出版商（属性是字符型）。

bookAbstract：图书摘要（属性是字符型）。

（3）orderForm 表

表名：orderForm。结构：如图 11.5 所示。

orderForm ：表

	字段名称	数据类型
🔑	orderNumber	数字
	logname	文本
	orderMess	文本
	sum	数字

图 11.5　orderForm 表

用途：存储订单信息，orderForm 表的主键是 orderNumber，各字段值的说明如下：

orderNumber：存储订单号（属性是 int 型）。
logname ：存储注册的用户名（属性是字符型）。
orderMess：订单信息（属性是字符型）。
sum：所订图书的价格总和（属性是单精度浮点型）。

2 数据库连接

系统采用 JDBC-ODBC 桥接器方式访问数据库。设置的数据源是 bookshop。选择"控制面板"→"管理工具"→"ODBC 数据源"（某些 Window 系统需选择"控制面板"→"性能和维护"→"管理工具"→"ODBC 数据源"），双击"ODBC 数据源"设置系统数据源，本系统设置的数据源名称是 bookshop，如图 11.6 所示。

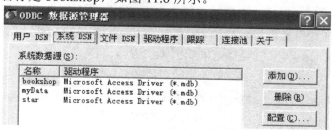

图 11.6 ODBC 数据源:bookshop

11.3 系统管理

本系统使用的 Web 服务目录是 chapter11，是在 Tomcat 安装目录的 webapps 目录下建立的 Web 服务目录。

需要在当前 Web 服务目录下建立如下的目录结构：
    chapter11\WEB-INF\tags

网站所涉及的 JSP 页面和图像均保存在目录 chapter11 中；Tag 文件保存在 chapter11\WEB-INF\tags 目录中。为了让 Tomcat 服务器启用上述目录，必须重新启动 Tomcat 服务器。

1. 页面管理

所有的页面将包括一个导航条，该导航条由注册、登录、浏览图书、浏览会员等组成。为了便于维护，其他页面通过使用 JSP 的<%@ include …%>标记将导航条文件 head.txt 嵌入到自己的页面。head.txt 保存在 Web 服务目录 chapter11 中。head.txt 的内容如下：

head.txt

```
<%@ page contentType="text/html;charset=GB2312" %>
<div align="center">
  <H2>网上书城</H2>
  <table   cellSpacing="1" cellPadding="1" width="760" align="center" border="0" >
   <tr valign="bottom">
    <td><A href="register.jsp"><font size=2>用户注册</font></A></td>
    <td><A href="login.jsp"><font size=2>用户登录</font></A></td>
    <td><A href="queryOrderForm.jsp"><font size=2>查看订单</font></A></td>
    <td><A href="lookPurchase.jsp"><font size=2>查看购物车</font></A></td>
```

```
            <td><A href="lookBook.jsp"><font size=2>浏览图书</font></A></td>
            <td><A href="findBook.jsp"><font size=2>图书查询</font></A></td>
            <td><A href="modifyRegister.jsp "><font size=2>修改注册信息</font></A></td>
            <td><A href="modifyPassword.jsp"><font size=2>修改密码</font></A></td>
            <td><A href="exitLogin.jsp"><font size=2>退出登录</font></A></td>
            <td><A href="index.jsp"><font size=2>返回主页</font></A></td>
          </tr>
        </Font>
    </table>
</div>
```

主页 index.jsp 由导航条、一句欢迎标语和一幅图片 welcome.jpg 组成，welcome.jpg 保存在 chapter11 中。

用户可以通过在浏览器的地址栏中键入"http://服务器 IP:8080/index.jsp"或"http://服务器 IP:8080/"访问该主页，主页运行效果如图 11.7 所示。

图 11.7 主页 index.jsp

index.jsp
```
    <%@ page contentType="text/html;charset=GB2312" %>
    <html>
    <head>
       <title>网上书城</title>
    </head>
    <body>
    <%@ include file="head.txt" %>
      <center>
      <h1><Font Size=4 color=green>欢迎浏览并订购图书</font></h1>
      <image src="welcome.jpg" width=300 height=200 ></image>
      </center>
    </body>
    </html>
```

2．Tag 文件的管理

在当前 Web 服务目录 chapter11 下建立如下目录结构：
    chapter11\WEB-INF\tags
网站所涉及 Tag 文件保存在 chapter11\WEB-INF\tags 目录中。

11.4 用户注册

该模块要求用户必须输入用户名、密码信息，否则不允许注册。用户的注册信息被存入数据库的 user 表中。

该模块由一个 JSP 页面 Register.jsp 和一个 Tag 文件 Register.tag 构成。Register.jsp 页面负责提交用户的注册信息到本页面，然后调用 Register.tag 文件。Register.tag 文件负责将用户提交的信息写入数据库的 user 表中。

1. JSP 页面

Register.jsp 页面负责提供输入注册信息界面，并显示注册反馈信息。该页面将用户提交的注册信息交给 Register.tag 文件，并显示 Tag 文件返回的有关注册是否成功的信息。Register.jsp 页面效果如图 11.8。

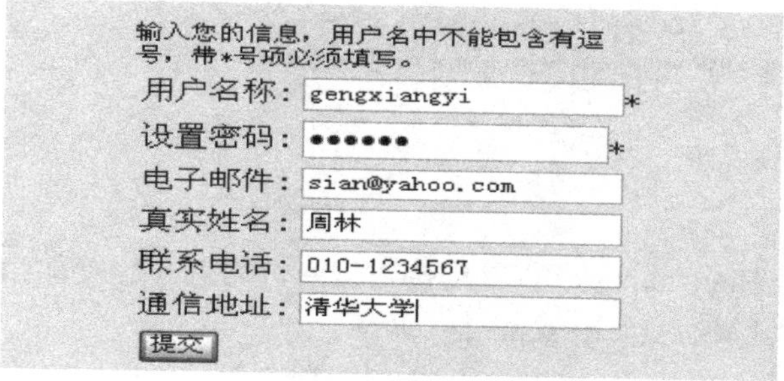

图 11.8 注册页面

Register.jsp

```
<%@ page contentType="text/html;charset=GB2312" %>
<HEAD><%@ include file="head.txt" %></HEAD>
<%@ taglib tagdir="/WEB-INF/tags" prefix="register"%>
<title>
    注册页面
</title>
<HTML>
<BODY bgcolor=cyan><Font size=2>
<CENTER>
<FORM action="" name=form>
<table>
    输入您的信息，用户名中不能包含有逗号，带*号项必须填写。
    <tr><td>用户名称:</td><td><Input type=text name="logname" >*</td></tr>
     <tr><td>设置密码:</td><td><Input type=password name="password">*</td></tr>
    <tr><td>电子邮件:</td><td><Input type=text name="email"></td></tr>
    <tr><td>真实姓名:</td><td><Input type=text name="realname"></td></tr>
    <tr><td>联系电话:</td><td><Input type=text name="phone"></td></tr>
    <tr><td>通信地址:</td><td><Input type=text name="address"></td></tr>
    <tr><td><Input type=submit name="g" value="提交"></td> </tr>
</table>
</Form>
```

```
</CENTER>
<%
    String logname=request.getParameter("logname");
    String password=request.getParameter("password");
    String email=request.getParameter("email");
    String realname=request.getParameter("realname");
    String phone=request.getParameter("phone");
    String address=request.getParameter("address");
%>
<register:Register logname="<%=logname%>"
                   password="<%=password%>"
                   email="<%=email%>"
                   realname="<%=realname%>"
                   phone="<%=phone%>"
                   address ="<%=address%>" />
<Center><P>返回的消息:<%=backMess %></Center>
</Body>
</HTML>
```

2. Tag 文件

Tag 文件的名字是 Register.tag，负责连接数据库，将用户提交的信息写入到 user 表，并返回有关注册是否成功的信息给 Register.jsp 页面。

Register.tag

```
<%@ tag import="java.sql.*" %>
<%@ tag pageEncoding="gb2312" %>
<%@ attribute name="logname" required="true" %>
<%@ attribute name="password" required="true" %>
<%@ attribute name="email" required="true" %>
<%@ attribute name="address" required="true" %>
<%@ attribute name="realname" required="true" %>
<%@ attribute name="phone" required="true" %>
<%@ variable name-given="backMess"  scope="AT_END" %>
<%   boolean boo=true;
     if(logname!=null){
         if(logname.contains(",")||logname.contains("，"))
             boo=false;
     }
     if(boo){
       try{   Class.forName("sun.jdbc.odbc.JdbcOdbcDriver");
       }
       catch(ClassNotFoundException e){
         out.print(e);//response.sendRedirect("error.jsp");
       }
       Connection con;
       Statement sql;
       ResultSet rs;
       String condition="INSERT INTO user VALUES";
```

```
            condition+="("+'"'+logname;
            condition+="','"+password;
            condition+="','"+phone;
            condition+="','"+email;
            condition+="','"+address;
            condition+="','"+realname+"')";
            try{
                    byte [] b=condition.getBytes("iso-8859-1");
                    condition=new String(b);
                    String uri= "jdbc:odbc:bookshop";
                    con=DriverManager.getConnection(uri,"","");
                    sql=con.createStatement();
                    sql.executeUpdate(condition);
                    con.close();
                    byte [] c=logname.getBytes("iso-8859-1");
                    logname=new String(c);
                    String mess=logname+"注册成功";
                    jspContext.setAttribute("backMess",mess);
                    con.close();
            }
            catch(Exception e){
                    jspContext.setAttribute("backMess","没有填写用户名或用户名已经被注册");
            }
      }
      else{
             jspContext.setAttribute("backMess","注册失败(用户名中不能有逗号)");
      }
%>
```

11.5 会员登录

用户在该模块输入曾注册的用户名和密码，该模块将对用户名和密码进行验证，如果输入的用户名或密码有错误，将提示用户输入的用户名或密码不正确。

该模块由一个 JSP 页面 Login.jsp 和一个 Tag 文件 Login.tag 构成。Login.jsp 负责提交用户的登录信息到本页面，然后页面调用 Login.tag 文件。Login.tag 负责验证用户名和密码是否正确，并返回登录是否成功的消息给 Login.jsp 页面。

1. JSP 页面

JSP 页面 Login.jsp 负责提交用户的登录信息到本页面，然后页面调用 Login.tag 文件，并负责显 Login.tag 文件的反馈信息，比如登录是否成功等，Login.jsp 页面效果如图 11.9 所示。

Login.jsp

```
<%@ page contentType="text/html;charset=GB2312" %>
<%@ taglib tagdir="/WEB-INF/tags" prefix="login"%>
<HEAD><%@ include file="head.txt" %></HEAD>
<title>
     登录页面
```

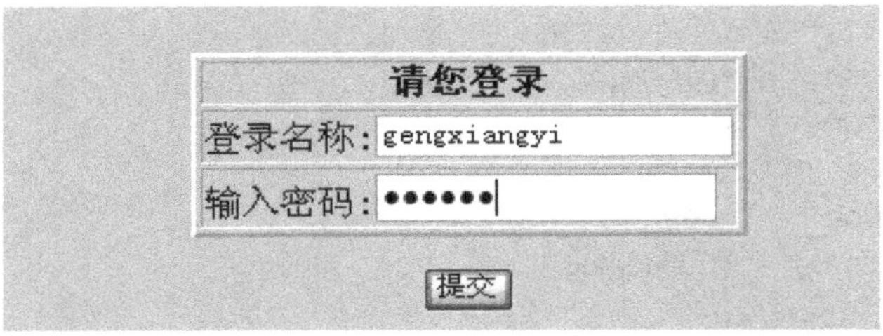

图 11.9 登录页面

```
</title>
<HTML>
<BODY bgcolor=pink><Font size=2><CENTER>
<BR><BR>
<table border=2>
<tr> <th>请您登录</th></tr>
<FORM action="" Method="post">
<tr><td>登录名称:<Input type=text name="logname"></td></tr>
<tr><td>输入密码:<Input type=password name="password"></td></tr>
</table>
<BR><Input type=submit name="g" value="提交">
</Form>
</CENTER>
<%
    String logname=request.getParameter("logname");
    if(logname==null){
       logname="";
    }
    String password=request.getParameter("password");
     if(password==null){
        password="";
    }
%>
 <login:Login logname="<%=logname%>"   password="<%=password%>" />
 <Center><P>返回的消息:<%=backMess %></Center>
</BODY>
</HTML>
```

2. Tag 文件

Login.tag 负责连接数据库，查询 user 表中的注册信息，以便验证用户名和密码是否正确，并返回登录是否成功的消息给 JSP 页面 Login.jsp。

Login.tag

```
<%@ tag import="java.sql.*" %>
<%@ tag pageEncoding="gb2312" %>
<%@ attribute name="logname" required="true" %>
<%@ attribute name="password" required="true" %>
<%@ variable name-given="backMess"   scope="AT_END" %>
<%
```

```java
byte [] a=logname.getBytes("iso-8859-1");
logname=new String(a);
byte [] b=password.getBytes("iso-8859-1");
password=new String(b);
String mess="";
try{   Class.forName("sun.jdbc.odbc.JdbcOdbcDriver");
}
catch(ClassNotFoundException e){
       out.print(e);
}
Connection con;
Statement sql;
ResultSet rs;
String loginMess=(String)session.getAttribute("logname");
if(loginMess==null){
     loginMess="*************";
}
String str=logname+","+password;
if(loginMess.equals(str)){
       mess=logname+"已经登录了";
}
else{
       String uri="jdbc:odbc:bookshop";
       boolean boo=(logname.length()>0)&&(password.length()>0);
       try{
           con=DriverManager.getConnection(uri,"","");
           String condition=
           "select * from user where logname = '"+
           logname+"' and password ='"+password+"'";
           sql=con.createStatement();
           if(boo) {
              rs=sql.executeQuery(condition);
              boolean m=rs.next();
              if(m==true) {
                  mess=logname+"登录成功";
                  str=logname+","+password;;
                  session.setAttribute("logname",str);
              }
              else {
                  mess="您输入的用户名"+logname+"不存在,或密码不般配";
              }
           }
           else
           {
               mess="还没有登录或您输入的用户名不存在、或密码不般配";
           }
           con.close();
```

```
            }
            catch(SQLException exp){
                mess="问题:"+exp;
            }
        }
        jspContext.setAttribute("backMess",mess);
%>
```

11.6 浏览图书信息

该模块由一个 JSP 页面 LookBook.jsp 和一个 Tag 文件 ShowBookByPage.tag 构成。LookBook.jsp 页面负责调用 ShowBookByPage.tag 文件，ShowBookByPage.tag 文件负责显示图书信息。

1．JSP 页面

LookBook.jsp 负责调用 ShowBookByPage.tag 文件，并将有关数据源、表的名称以及需要显示的页码等信息传递给该 Tag 文件，然后显示 Tag 文件返回的有关信息。LookBook.jsp 页面的效果如图 11.10 所示。

图 11.10　浏览图书页面

LookBook.jsp
```
<%@ page contentType="text/html;charset=GB2312" %>
<%@ taglib tagdir="/WEB-INF/tags" prefix="showBookByPage"%>
<%@ include file="head.txt" %></HEAD>
<HTML>
<Body bgcolor=yellow><center>
 <%
    String number=request.getParameter("page");
    if(number==null){
        number="1";
    }
%>
<BR>每页最多显示 2 本图书
<showBookByPage:ShowBookByPage dataSource="bookshop"   tableName="bookForm"
                               bookAmountInPage="2"   zuduanAmount="6"
```

```
                          page="<%=number%>"/>
<BR>共有<%=pageAllCount%>页,当前显示第<%=showPage%>页
<BR><%=giveResult%>
<%
    int m=showPage.intValue();
%>
<a href ="lookBook.jsp?page=<%=m+1%>">下一页</a>
<a href ="lookBook.jsp?page=<%=m-1%>">上一页</a>
<form action="">
  输入页码：<Input type=text name="page" >
  <Input type=submit value="提交">
</form>
</BODY>
</HTML>
```

2. Tag 文件

ShowBookByPage.tag 负责连接数据库，查询 bookForm 表，并将查询到的图书信息反馈给 JSP 页面 LookBook.jsp。

ShowBookByPage.tag

```
<%@ tag import="java.sql.*" %>
<%@ tag import="com.sun.rowset.*" %>
<%@ tag pageEncoding="gb2312" %>
<%@ attribute name="dataSource" required="true" %>
<%@ attribute name="tableName" required="true" %>
<%@ attribute name="bookAmountInPage" required="true" %>
<%@ attribute name="page" required="true" %>
<%@ attribute name="zuduanAmount" required="true" %>
<%@ variable name-given="showPage" variable-class="java.lang.Integer"
      scope="AT_END" %>
<%@ variable name-given="pageAllCount"
      variable-class="java.lang.Integer" scope="AT_END" %>
<%@ variable name-given="giveResult"
      variable-class="java.lang.StringBuffer" scope="AT_END" %>
<%
    try{  Class.forName("sun.jdbc.odbc.JdbcOdbcDriver");  }
    catch(ClassNotFoundException e){  out.print(e);  }
    Connection con;
    Statement sql;
    ResultSet rs;
    int pageSize=Integer.parseInt(bookAmountInPage);   //每页显示的记录数
    int allPages=0;                                    //分页后的总页数
    int show=Integer.parseInt(page);                   //当前显示页
    StringBuffer presentPageResult;                    //当前页上的内容
    CachedRowSetImpl rowSet;
    presentPageResult=new StringBuffer();
    String uri="jdbc:odbc:"+dataSource;
    try{    con=DriverManager.getConnection(uri,"","");
```

```java
sql=con.createStatement(ResultSet.TYPE_SCROLL_SENSITIVE,
                       ResultSet.CONCUR_READ_ONLY);
String s= "select * from "+tableName;
rs=sql.executeQuery(s);
rowSet=new CachedRowSetImpl();              //创建行集对象
rowSet.populate(rs);
con.close();                                //关闭连接
rowSet.last();
int m=rowSet.getRow();                      //总行数
int n=pageSize;
allPages=((m%n)==0)?(m/n):(m/n+1);
int p=Integer.parseInt(page);
if(p>allPages)
   p=1;
if(p<=0)
   p=allPages;
jspContext.setAttribute("showPage",new Integer(p));
jspContext.setAttribute("pageAllCount",new Integer(allPages));
presentPageResult.append("<table border=1>");
presentPageResult.append("<tr>");
presentPageResult.append("<th>封面</td>");
presentPageResult.append("<th>ISBN</td>");
presentPageResult.append("<th>图书名称</td>");
presentPageResult.append("<th>作者</td>");
presentPageResult.append("<th>价格</td>");
presentPageResult.append("<th>出版社</td>");
presentPageResult.append("</tr>");
rowSet.absolute((p-1)*pageSize+1);
int 字段个数=6;
字段个数=Integer.parseInt(zuduanAmount);
for(int i=1;i<=pageSize;i++){
    presentPageResult.append("<tr>");
    String bookISBN="";
    for(int k=1;k<=字段个数;k++) {
        if(k==1){
            String bookPic="<image src="+rowSet.getString(k)+
            " width=70 height=100/>";
            presentPageResult.append("<td>"+bookPic+"</td>");
        }
        else if(k==2) {
            bookISBN=rowSet.getString(k);
            String bookISBNLink="<a href=\"lookBookAbstract.jsp?bookISBN="+
            bookISBN+"\">"+bookISBN+"</a>";
            presentPageResult.append("<td>"+bookISBNLink+"</td>");
        }
          else if(k==3) {
            String bookName=rowSet.getString(k);
```

```
                                String bookNameLink="<a href=\"lookBookAbstract.jsp?bookISBN="+
                                    bookISBN+"\">"+bookName+"</a>";
                                presentPageResult.append("<td>"+bookNameLink+"</td>");
                            }
                            else {
                                presentPageResult.append("<td>"+rowSet.getString(k)+"</td>");
                            }

                        }
                        String buy="<a href=\"lookPurchase.jsp?buyISBN="+bookISBN+"\">购买</a>";
                        presentPageResult.append("<td>"+buy+"</td>");
                        presentPageResult.append("</tr>");
                        boolean boo=rowSet.next();
                        if(boo==false) break;
                    }
                    presentPageResult.append("</table>");
                    jspContext.setAttribute("giveResult",presentPageResult);
                    con.close();
            }
            catch(SQLException exp){
                    jspContext.setAttribute("showPage",new Integer(1));
                    jspContext.setAttribute("pageAllCount",new Integer(1));
                    jspContext.setAttribute("giveResult",new StringBuffer(""+exp));
            }
%>
```

11.7 查询图书

该模块由一个 JSP 页面 FindBook.jsp 和一个 Tag 文件 FindBook.tag 构成。FindBook.jsp 页面负责调用 FindBook.tag 文件，FindBook.tag 文件负责显示图书信息。

1. JSP 页面

FindBook.jsp 页面负责调用 FindBook.tag 文件，并将有关数据源、表的名称以及 ISBN、作者姓名或图书名称等信息传递给该 Tag 文件，然后显示 Tag 文件返回的有关信息。FindBook.jsp 页面的效果如图 11.11 所示。

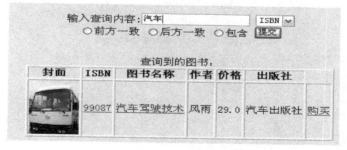

图 11.11 图书查询页面

FindBook.jsp

```
<%@ page contentType="text/html;charset=GB2312" %>
<%@ taglib tagdir="/WEB-INF/tags" prefix="findBook"%>
<%@ include file="head.txt" %></HEAD>
<HTML>
<Body bgcolor=cyan><center>
 <form action="">
    输入查询内容:<Input type=text name="findContent" value="java">
       <Select name="condition" size=1>
          <Option Selected value="bookISBN">ISBN
          <Option value="bookName">书名
          <Option value="bookAuthor">作者
       </Select>
       <Br>
       <INPUT type="radio" name="findMethod" value="start">前方一致
       <INPUT type="radio" name="findMethod" value="end">后方一致
       <INPUT type="radio" name="findMethod" value="contains">包含
         <Input type=submit value="提交">
   </form>
<%
    String findContent = request.getParameter("findContent");
    String condition = request.getParameter("condition");
    String findMethod = request.getParameter("findMethod");
    if(findContent==null){
        findContent="";
    }
    if(condition==null){
        condition="";
    }
    if(findMethod==null){
        findMethod="";
    }
%>
  <BR>查询到的图书:
  <findBook:FindBook dataSource="bookshop"
                 tableName="bookForm"
                 findContent="<%=findContent%>"
                 condition="<%=condition%>"
                 findMethod="<%=findMethod%>"/>
   <BR><%=giveResult%>
  </form>
</BODY>
</HTML>
```

2. Tag 文件

FindBook.tag 文件负责连接数据库，查询 bookForm 表，并将查询到的图书信息反馈给 JSP 页面 FindBook.jsp。

FindBook.tag

```jsp
<%@ tag import="java.sql.*" %>
<%@ tag pageEncoding="gb2312" %>
<%@ attribute name="dataSource" required="true" %>
<%@ attribute name="tableName" required="true" %>
<%@ attribute name="findContent" required="true" %>
<%@ attribute name="condition" required="true" %>
<%@ attribute name="findMethod" required="true" %>
<%@ variable name-given="giveResult" variable-class=
"java.lang.StringBuffer" scope="AT_END" %>
<%
    byte b[]=findContent.getBytes("iso-8859-1");
    findContent=new String(b);
    try{   Class.forName("sun.jdbc.odbc.JdbcOdbcDriver");   }
    catch(ClassNotFoundException e){   out.print(e);   }
    Connection con;
    Statement sql;
    ResultSet rs;
    StringBuffer queryResult=new StringBuffer();       //查询结果
    String uri="jdbc:odbc:"+dataSource;
    try{     con=DriverManager.getConnection(uri,"","");
         sql=con.createStatement();
         String s="";
         if(findMethod.equals("start"))
             s= "select * from "+tableName+" where "+
             condition+" Like'"+findContent+"%'";
         if(findMethod.equals("end"))
             s= "select * from "+tableName+" where "+
             condition+" Like'%"+findContent+"'";
         if(findMethod.equals("contains"))
             s= "select * from "+tableName+" where "+
             condition+" Like'%"+findContent+"%'";
         rs=sql.executeQuery(s);
         queryResult.append("<table border=1>");
         queryResult.append("<tr>");
         queryResult.append("<th>封面</td>");
         queryResult.append("<th>ISBN</td>");
         queryResult.append("<th>图书名称</td>");
         queryResult.append("<th>作者</td>");
         queryResult.append("<th>价格</td>");
         queryResult.append("<th>出版社</td>");
         queryResult.append("</tr>");
         int 字段个数=6;
         while(rs.next()){
             queryResult.append("<tr>");
             String bookISBN="";
             for(int k=1;k<=字段个数;k++) {
                 if(k==1){
```

```
                String bookPic=
                   "<image src="+rs.getString(k)+" width=70 height=100/>";
                      queryResult.append("<td>"+bookPic+"</td>");
                }
                else if(k==2) {
                      bookISBN=rs.getString(k);
                      String bookISBNLink=
                   "<a href=\"lookBookAbstract.jsp?bookISBN="+
                      bookISBN+"\">"+bookISBN+"</a>";
                      queryResult.append("<td>"+bookISBNLink+"</td>");
                }
                else if(k==3) {
                      String bookName=rs.getString(k);
                      String bookNameLink=
                   "<a href=\"lookBookAbstract.jsp?bookISBN="+
                      bookISBN+"\">"+bookName+"</a>";
                      queryResult.append("<td>"+bookNameLink+"</td>");
                }
                else {
                      queryResult.append("<td>"+rs.getString(k)+"</td>");
                }
            }
            String buy="<a href=\"lookPurchase.jsp?buyISBN="+
                      bookISBN+"\">购买</a>";
            queryResult.append("<td>"+buy+"</td>");
        }
        queryResult.append("</table>");
        jspContext.setAttribute("giveResult",queryResult);
        con.close();
    }
    catch(SQLException exp){
        jspContext.setAttribute("giveResult",new StringBuffer("请给出查询条件"));
    }
%>
```

11.8 查看购物车

该模块由一个 JSP 页面 LookPurchase.jsp 和一个 Tag 文件 LookPurchase.tag 构成。LookPurchase.jsp 页面负责调用 LookPurchase.tag 文件，LookPurchase.tag 文件负责显示用户购物车（session 对象）中的图书。

1. JSP 页面

LookPurchase.jsp 负责将用户购买的图书添加到用户的购物车（session 对象），并可以根据用户的选择从购物车中删除曾添加到购物车中的图书。LookPurchase.jsp 页面负责调用 LookPurchase.tag 文件，并显示 Tag 文件返回的有关信息。用户在 LookPurchase.jsp 页面可以确定是否生成订单。LookPurchase.jsp 页面的效果如图 11.12 所示。

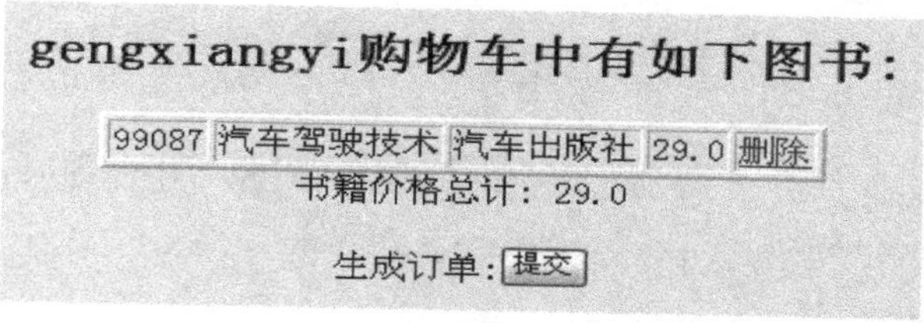

图 11.12　购物车页面

LookPurchase.jsp

```jsp
<%@ page contentType="text/html;charset=GB2312" %>
<%@ taglib tagdir="/WEB-INF/tags" prefix="lookPurchase"%>
<HEAD><%@ include file="head.txt" %></HEAD>
<title>
    查看购物车
</title>
<html>
<body bgcolor=cyan><center>
<%
    boolean isAdd=false;
    String logname=(String)session.getAttribute("logname");
    if(logname!=null){
       int m=logname.indexOf(",");
       logname=logname.substring(0,m);
       isAdd=true;
    }
    else{
        response.sendRedirect("login.jsp");
    }
    String buyISBN=request.getParameter("buyISBN");
    if((buyISBN!=null)&&isAdd){
        session.setAttribute(buyISBN+","+logname,buyISBN);
    }
    String deletedISBN=request.getParameter("deletedISBN");
    if((deletedISBN!=null)&&isAdd){
        session.removeAttribute(deletedISBN+","+logname);
    }
%>
    <lookPurchase:LookPurchase logname="<%=logname%>"/>
    <h2><%=logname%>购物车中有如下图书:</h2>
    <%=giveResult%>
    书籍价格总计:
    <%=price%>
    <form action="previewOrderForm.jsp">
       生成订单:<Input type=submit name="g" value="提交">
    </form>
</center>
</body>
```

</html>

2. Tag 文件

LookPurchase.tag 文件负责显示用户购物车（session 对象）中的图书，并计算出购物车中图书的总价，然后将这些信息返回给 LookPurchase.jsp 页面。

LookPurchase.tag

```jsp
<%@ tag import="java.sql.*" %>
<%@ tag import="java.util.*" %>
<%@ tag pageEncoding="gb2312" %>
<%@ attribute name="logname" required="true" %>
<%@ variable name-given="giveResult" variable-class=
"java.lang.StringBuffer" scope="AT_END" %>
<%@ variable name-given="price" variable-class=
"java.lang.Float" scope="AT_END" %>
<%
        float totalPrice=0;
        String bookISBN;
        String bookName;
        String bookPublish;
        float bookPrice;
        String uri="jdbc:odbc:bookshop";
        Connection con;
        Statement sql;
        ResultSet rs;
        StringBuffer str=new StringBuffer();
        try{   Class.forName("sun.jdbc.odbc.JdbcOdbcDriver");
        }
        catch(ClassNotFoundException e){
            str.append(e);
        }
        Enumeration keys=session.getAttributeNames();
        str.append("<table border=2>");
        while(keys.hasMoreElements()) {
            String key=(String)keys.nextElement();
            boolean isTrue=(!(key.equals("logname")))&&(key.endsWith(logname));
            if(isTrue){
                bookISBN=(String)session.getAttribute(key);
                String sqlStatement=
                "select * from bookForm where bookISBN = '"+bookISBN+"'" ;
                try{
                    con=DriverManager.getConnection(uri,"","");
                    sql=con.createStatement();
                    rs=sql.executeQuery(sqlStatement);
                    while(rs.next()){
                        bookISBN=rs.getString("bookISBN");
                        bookName=rs.getString("bookName");
                        bookPublish=rs.getString("bookPublish");
```

```
                    bookPrice=rs.getFloat("bookPrice");
                    totalPrice=totalPrice+bookPrice;
                    str.append("<tr>");
                    str.append("<td>"+bookISBN+"</td>");
                    str.append("<td>"+bookName+"</td>");
                    str.append("<td>"+bookPublish+"</td>");
                    str.append("<td>"+bookPrice+"</td>");
                    String del=
                    "<a href=\"lookPurchase.jsp?deletedISBN="+bookISBN+"\">删除</a>";
                    str.append("<td>"+del+"</td>");
                    str.append("</tr>");
                }
                con.close();
            }
            catch(SQLException exp){
                str.append(exp);
            }
        }
    }
    str.append("</table>");
    jspContext.setAttribute("giveResult",str);
    jspContext.setAttribute("price",new Float(totalPrice));
%>
```

10.9 订单预览

该模块由一个 JSP 页面 PreviewOrderForm.jsp 和一个 Tag 文件 PreviewOrderForm.tag 构成。PreviewOrderForm.jsp 负责调用 PreviewOrderForm.tag 文件，并显示 PreviewOrderForm.tag 文件返回的待确定的订单。

1. JSP 页面

PreviewOrderForm.jsp 页面负责调用 LookPurchasek.tag 文件，并显示其返回的订单。用户在 PreviewOrderForm.jsp 页面可以选择是否确定订单。PreviewOrderForm.jsp 页面的效果如图 11.13 所示。

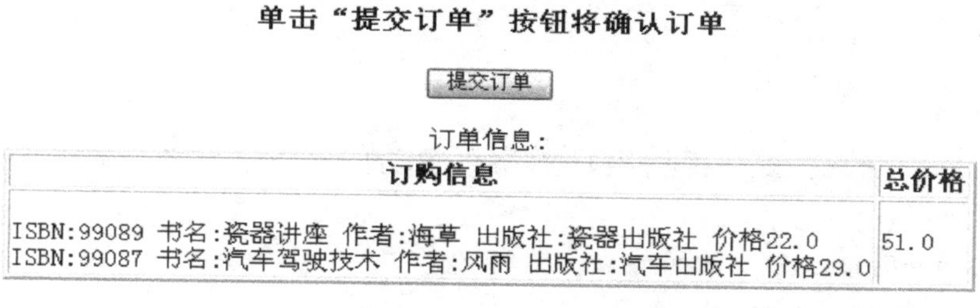

图 11.13 订单预览页面

PreviewOrderForm.jsp

```
<%@ page contentType="text/html;charset=GB2312" %>
```

```jsp
<%@ taglib tagdir="/WEB-INF/tags" prefix="previewOrderForm" %>
<HEAD><%@ include file="head.txt" %></HEAD>
<title>
    当前订单(预览)
</title>
<%
    String logname=(String)session.getAttribute("logname");
    if(logname!=null){
      int m=logname.indexOf(",");
      logname=logname.substring(0,m);
%>
     <previewOrderForm:PreviewOrderForm logname="<%=logname%>" />
     <HTML>
     <Body ><center>
     <h3>单击"提交订单"按钮将确认订单</h3>
     <form action="makeBookForm.jsp">
          <Input type=hidden name="confirm" value="buy">
          <Input type=hidden name="orderContent" value="<%= giveResult %>">
          <Input type=hidden name="totalPrice" value="<%= totalPrice %>">
          <center> <Input type=submit name="g" value="提交订单"></center>
     </form>
     订单信息:<br>
     <table border=2>
         <tr><th>订购信息</th>
             <th>总价格</th>
         </tr>
         <tr>
            <td><%= giveResult %></td>
            <td><%=totalPrice %> </td>
         </tr>
     </center>
     </BODY>
     </HTML>
<%  }
    else{
       response.sendRedirect("login.jsp");
    }
%>
```

2. Tag 文件

PreviewOrderForm.tag 文件根据用户购物车中的图书生成订单,然后将订单返回给 PreviewOrderForm.jsp 页面。

PreviewOrderForm.tag

```jsp
<%@ tag import="java.util.*" %>
<%@ tag import="java.sql.*" %>
<%@ tag pageEncoding="gb2312" %>
<%@ attribute name="logname" required="true" %>
<%@ variable name-given="giveResult" variable-class="java.lang.StringBuffer" scope="AT_END" %>
```

```jsp
<%@ variable name-given="totalPrice" variable-class="java.lang.Float" scope="AT_END" %>
<%
    try{   Class.forName("sun.jdbc.odbc.JdbcOdbcDriver");   }
    catch(ClassNotFoundException e){   out.print(e);   }
    Connection con;
    Statement sql;
    ResultSet rs;
    StringBuffer orderMess=new StringBuffer();
    String uri="jdbc:odbc:bookshop";
    try{
        con=DriverManager.getConnection(uri,"","");
        sql=con.createStatement();
        Enumeration keys=session.getAttributeNames();
        float sum=0;
        while(keys.hasMoreElements()){
            String key=(String)keys.nextElement();
            boolean isTrue=(!(key.equals("logname")))&&(key.endsWith(logname));
            if(isTrue) {
                String bookISBN=(String)session.getAttribute(key);
                String sqlStatement=
                "select * from bookForm where bookISBN = '"+bookISBN+"'" ;
                rs=sql.executeQuery(sqlStatement);
                while(rs.next()){
                    bookISBN=rs.getString("bookISBN");
                    String bookName=rs.getString("bookName");
                    String bookAuthor=rs.getString("bookAuthor");
                    String bookPublish=rs.getString("bookPublish");
                    float bookPrice=rs.getInt("bookPrice");
                    sum=sum+bookPrice;
                    orderMess.append("<br>ISBN:"+bookISBN+" 书名:"+bookName+
                    " 作者:"+bookAuthor+" 出版社:"+bookPublish+" 价格"+bookPrice);
                }
            }
        }
        jspContext.setAttribute("giveResult",orderMess);
        jspContext.setAttribute("totalPrice",new Float(sum));
    }
    catch(SQLException exp){
        jspContext.setAttribute("giveResult",new StringBuffer("没有订单"));
        jspContext.setAttribute("totalPrice",new Float(-1));
    }
%>
```

11.10 确认订单

该模块由一个 JSP 页面 MakeOrderForm.jsp 和一个 Tag 文件 MakeOrderForm.tag 构成。MakeOrderForm.jsp 负责调用 MakeOrderForm.tag 文件，并显示 MakeOrderFormk.tag 返回的确认

的订单。

1. JSP 页面

MakeOrderForm.jsp 页面负责调用 MakeOrderForm.tag 文件，并将订购图书的有关信息传递给该 Tag 文件。MakeOrderForm.jsp 页面的效果如图 11.14 所示。

付款后发货

gengxiangyi当前的订单号:2014
订单信息:

订单号	订单内容	总价格
2014	ISBN:99089 书名:瓷器讲座 作者:海草 出版社:瓷器出版社 价格22.0 ISBN:99087 书名:汽车驾驶技术 作者:风雨 出版社:汽车出版社 价格29.0	51.0

图 11.14 确认订单页面

MakeOrderForm.jsp

```jsp
<%@ page contentType="text/html;charset=GB2312" %>
<%@ taglib tagdir="/WEB-INF/tags" prefix="makeBookForm" %>
<HEAD><%@ include file="head.txt" %></HEAD>
<title>
    订单确认
</title>
<%
    String logname=(String)session.getAttribute("logname");
    if(logname==null){
        response.sendRedirect("login.jsp");
    }
    else{
        int m=logname.indexOf(",");
        logname=logname.substring(0,m);
    }
    String confirm=request.getParameter("confirm");
    String orderContent=request.getParameter("orderContent");
    String totalPrice=request.getParameter("totalPrice");
    if(confirm==null){
        confirm="";
    }
    if(orderContent==null){
        orderContent="";
    }
    if(totalPrice==null){
        totalPrice="0";
    }
    byte[] a=orderContent.getBytes("iso-8859-1");
    orderContent=new String(a);
    a=totalPrice.getBytes("iso-8859-1");
    totalPrice=new String(a);
```

```
                if(confirm.equals("buy")){
%>              <makeBookForm:MakeBookForm logname="<%=logname%>"
   orderContent="<%=orderContent%>"
                                        totalPrice="<%=totalPrice%>" />
        <HTML>
        <Body ><center>
        <h3>付款后发货</h3>
        <%=logname%>当前的订单号:<%=dingdanNumber%><br>
        订单信息:<br>
        <%= giveResult %>
        </center>
        </BODY>
        </HTML>
<%   }
%>
```

2．Tag 文件

MakeOrderForm.tag 文件负责连接数据库，将订单写入到数据库中的 orderForm 表，达到确定订单的目的，然后将订单信息返回给 MakeOrderForm.jsp 页面。

MakeOrderForm.tag

```
<%@ tag import="java.sql.*" %>
<%@ tag import="java.util.*" %>
<%@ tag pageEncoding="gb2312" %>
<%@ attribute name="logname" required="true" %>
<%@ attribute name="orderContent" required="true" %>
<%@ attribute name="totalPrice" required="true" %>
<%@ variable name-given="giveResult" variable-class=
"java.lang.StringBuffer" scope="AT_END" %>
<%@ variable name-given="dingdanNumber" variable-class=
"java.lang.Long" scope="AT_END" %>
<%
    String user=(String)session.getAttribute("logname");
    if(user==null){
        response.sendRedirect("login.jsp");
    }
    float sum=Float.parseFloat(totalPrice);
    try{  Class.forName("sun.jdbc.odbc.JdbcOdbcDriver");
    }
    catch(ClassNotFoundException e){
        out.print(e);
    }
    Connection con;
    Statement sql;
    ResultSet rs;
    String uri="jdbc:odbc:bookshop";
    int orderNumber=2010;
    int max=orderNumber;
```

```
String sqlStatement="";
try{
        con=DriverManager.getConnection(uri,"","");
        sql=con.createStatement(ResultSet.TYPE_SCROLL_SENSITIVE,
        ResultSet.CONCUR_READ_ONLY);
        rs=sql.executeQuery("SELECT * FROM  orderForm");
        while(rs.next()){
            int n=rs.getInt("orderNumber");
            if(n>=max)
                max=n;
        }
        orderNumber=max+1;
        sqlStatement="INSERT INTO orderForm VALUES ("+
        orderNumber+",'"+logname+"','"+orderContent+"',"+sum+")";
        sql.executeUpdate(sqlStatement);
        StringBuffer strMess=new StringBuffer();
        sqlStatement="select * from orderForm where orderNumber = "+orderNumber ;
        rs=sql.executeQuery(sqlStatement);
        strMess.append("<table border=2>");
        strMess.append("<tr>");
        strMess.append("<th>订单号</th>");
        strMess.append("<th>订单内容</th>");
        strMess.append("<th>总价格</th>");
        strMess.append("</tr>");
        while(rs.next()){
            String idNumber=rs.getString("orderNumber");
            String orderMess=rs.getString("orderMess");
            float priceSum=rs.getFloat("sum");
            strMess.append("<tr>");
                strMess.append("<td>"+idNumber+"</td>");
                strMess.append("<td>"+orderMess+"</td>");
                strMess.append("<td>"+priceSum+"</td>");
            strMess.append("</tr>");
        }
        strMess.append("</table>");
        jspContext.setAttribute("giveResult",strMess);
        jspContext.setAttribute("dingdanNumber",new Long(orderNumber));
        con.close();
}
catch(SQLException exp){
        jspContext.setAttribute("giveResult",new StringBuffer(""+exp));
        jspContext.setAttribute("dingdanNumber",new Long(-1));
}
%>
```

11.11 查询订单

该模块由一个 JSP 页面 QueryOrderForm..jsp 和一个 Tag 文件 QueryOrderForm.tag 构成。

QueryOrderForm.jsp 负责调用 QueryOrderForm.tag 文件，并显示该 Tag 文件返回的订单信息。

1. JSP 页面

QueryOrderForm.jsp 页面负责调用 QueryOrderForm.tag 文件，并将用户登录的用户名传递给该 Tag 文件。QueryOrderForm.jsp 页面的效果如图 11.15 所示。

gengxiangyi全部订单：

订单号	订单用户	订单信息	总价格
2013	gengxiangyi	ISBN:99087 书名:汽车驾驶技术 作者:风雨 出版社:汽车出版社 价格29.0	29.0
2014	gengxiangyi	ISBN:99089 书名:瓷器讲座 作者:海草 出版社:瓷器出版社 价格22.0 ISBN:99087 书名:汽车驾驶技术 作者:风雨 出版社:汽车出版社 价格29.0	51.0

图 11.15　查询订单页面

QueryOrderForm.jsp

```
<%@ page contentType="text/html;charset=GB2312" %>
<%@ taglib tagdir="/WEB-INF/tags" prefix="queryOrderForm" %>
<HEAD><%@ include file="head.txt" %></HEAD>
<title>
    查询订单
</title>
<%
    String logname=(String)session.getAttribute("logname");
    if(logname==null){
        response.sendRedirect("login.jsp");
    }
    else{
        int m=logname.indexOf(",");
        logname=logname.substring(0,m);
    }

%>
<queryOrderForm:QueryOrderForm logname="<%=logname%>" />
<HTML>
 <Body ><center>
    <h3><%=logname%>全部订单:</h3><br>
<%= giveResult %>
    </center>
    </BODY>
    </HTML>
```

2. Tag 文件

QueryOrderForm.tag 文件负责连接数据库，查询 orderForm 表，然后将订单信息返回给 QueryOrderForm.jsp 页面。

QueryOrderForm.tag

```jsp
<%@ tag import="java.sql.*" %>
<%@ tag pageEncoding="gb2312" %>
<%@ attribute name="logname" required="true" %>
<%@ variable name-given="giveResult" variable-class=
              "java.lang.StringBuffer" scope="AT_END" %>
<%
    try{ Class.forName("sun.jdbc.odbc.JdbcOdbcDriver"); }
    catch(ClassNotFoundException e){ out.print(e); }
    StringBuffer str=new StringBuffer();
    Connection con;
    Statement sql;
    ResultSet rs;
    String uri="jdbc:odbc:bookshop";
    try{   con=DriverManager.getConnection(uri,"","");
           sql=con.createStatement();
           String s= "select * from orderForm where logname= '"+logname+"'" ;
           rs=sql.executeQuery(s);
           str.append("<table border=1>");
           str.append("<tr>");
           str.append("<th>订单号</th>");
           str.append("<th>订单用户</th>");
           str.append("<th>订单信息</th>");
           str.append("<th>总价格</th>");
           str.append("</tr>");
           while(rs.next()){
                str.append("<tr>");
                str.append("<td>"+rs.getString(1)+"</td>");
                str.append("<td>"+rs.getString(2)+"</td>");
                str.append("<td>"+rs.getString(3)+"</td>");
                str.append("<td>"+rs.getString(4)+"</td>");
                str.append("</tr>");
           }
           str.append("</table>");
           jspContext.setAttribute("giveResult",str);
    }
    catch(SQLException exp){
           jspContext.setAttribute("giveResult",new StringBuffer(""+exp));
    }
%>
```

11.12 查看图书摘要

该模块由一个 JSP 页面 LookBookAbstract.jsp 和一个 Tag 文件 BookAbstract.tag 构成。LookBookAbstract.jsp 负责调用 BookAbstract.tag 文件，并显示 BookAbstract.tag 返回的图书摘要。

1. JSP 页面

LookBookAbstract.jsp 页面负责调用 BookAbstract.tag 文件，并将图书的 ISBN 传递给该 Tag 文件。LookBookAbstract.jsp 页面的效果如图 11.16 所示。

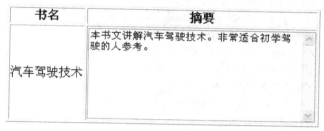

图 11.16　查看图书摘要

LookBookAbstract.jsp

```
<%@ page contentType="text/html;charset=GB2312" %>
<%@ taglib tagdir="/WEB-INF/tags" prefix="bookAbstract" %>
<HEAD><%@ include file="head.txt" %></HEAD>
<title>
    书的摘要
</title>
<%
    String bookISBN=request.getParameter("bookISBN");
%>
    <bookAbstract:BookAbstract bookISBN="<%=bookISBN%>" />
<HTML>
  <Body ><center>
   <%= giveResult %>
   </center>
   </BODY>
   </HTML>
```

2. Tag 文件

BookAbstract.tag 文件负责连接数据库，查询 orderForm 表，然后将图书摘要返回给 LookBookAbstract.jsp 页面。

BookAbstract.tag

```
<%@ tag import="java.sql.*" %>
<%@ tag pageEncoding="gb2312" %>
<%@ attribute name="bookISBN" required="true" %>
<%@ variable name-given="giveResult" variable-class=
             "java.lang.StringBuffer" scope="AT_END" %>
<%
    try{  Class.forName("sun.jdbc.odbc.JdbcOdbcDriver");  }
    catch(ClassNotFoundException e){   out.print(e);   }
    StringBuffer str=new StringBuffer();
    Connection con;
    Statement sql;
```

```
            ResultSet rs;
            String uri="jdbc:odbc:bookshop";
       try{     con=DriverManager.getConnection(uri,"","");
                sql=con.createStatement();
                String s= "select * from bookForm where bookISBN= '"+bookISBN+"'" ;
                rs=sql.executeQuery(s);
                str.append("<table border=1>");
                str.append("<tr>");
                str.append("<th>书名</th>");
                str.append("<th>摘要</th>");
                str.append("</tr>");
                while(rs.next()){
                    str.append("<tr>");
                    str.append("<td>"+rs.getString("bookName")+"</td>");
                    str.append("<td><TextArea Rows=8 Cols=40>"+
                    rs.getString("bookAbstract")+"</TextArea></td>");
                    str.append("</tr>");
                }
                str.append("</table>");
                jspContext.setAttribute("giveResult",str);
                con.close();
           }
           catch(SQLException exp){
                jspContext.setAttribute("giveResult",new StringBuffer(""+exp));
           }
      %>
```

11.13 修改密码

该模块由一个 JSP 页面 ModifyPassword.jsp 和一个 Tag 文件 ModifyPassword.tag 构成。ModifyPassword.jsp 负责调用 Tag 文件 ModifyPassword.tag 文件，并显示 ModifyPassword.tag 返回的有关修改密码是否成功的信息。

1. JSP 页面

ModifyPassword.jsp 页面负责调用 ModifyPassword.tag 文件，并将用户名、当前密码和新密码传递给该 Tag 文件。ModifyPassword.jsp 页面的效果如图 11.17 所示。

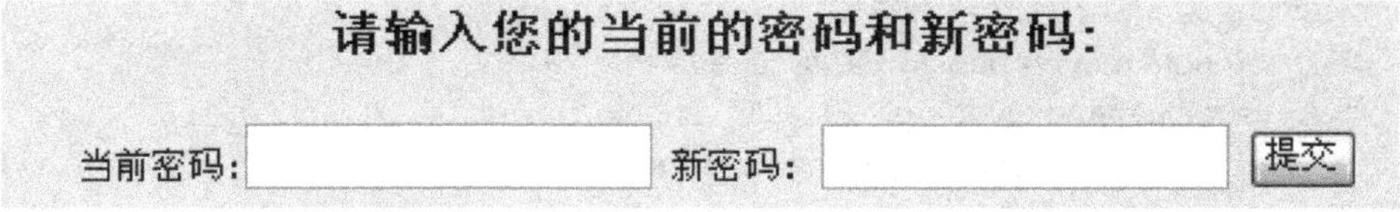

图 11.17 修改密码

ModifyPassword.jsp
```
<%@ page contentType="text/html;charset=GB2312" %>
<%@ taglib tagdir="/WEB-INF/tags" prefix="modifyPassword" %>
<HEAD><%@ include file="head.txt" %></HEAD>
```

```html
<HTML>
<BODY bgcolor=cyan>
<Font size=2>
<CENTER>
<h3>请输入您的当前的密码和新密码:</h3>
<FORM action="" Method="post">
    当前密码:<Input type=password name="oldPassword">
    新密码: <Input type=password name="newPassword">
        <Input type=submit name="g" value="提交">
</Form>
</CENTER>
</BODY>
</HTML>
<%
    boolean isModify=false;
    String logname=(String)session.getAttribute("logname");
    if(logname!=null){
      int m=logname.indexOf(",");
      logname=logname.substring(0,m);
      isModify=true;
    }
    else{
       response.sendRedirect("login.jsp");
    }
    String oldPassword=request.getParameter("oldPassword");
    String newPassword=request.getParameter("newPassword");
    boolean ok=oldPassword!=null&&newPassword!=null;
    if(ok&&isModify){
%>     <modifyPassword:ModifyPassword logname="<%=logname%>"
                         oldPassword="<%=oldPassword%>"
                         newPassword="<%=newPassword%>"/>
     <center><h2><%= giveResult %></h2></center>
<% }
%>
```

2. Tag 文件

ModifyPassword.tag 文件负责连接数据库，查询、更新 user 表，然后将密码是否更新成功的信息返回给 ModifyPasswod.jsp 页面。

ModifyPassword.tag

```
<%@ tag import="java.sql.*" %>
<%@ tag pageEncoding="gb2312" %>
<%@ attribute name="logname" required="true" %>
<%@ attribute name="oldPassword" required="true" %>
<%@ attribute name="newPassword" required="true" %>
<%@ variable name-given="giveResult" variable-class="java.lang.StringBuffer" scope="AT_END" %>
<%
    try{   Class.forName("sun.jdbc.odbc.JdbcOdbcDriver");   }
```

```
            catch(ClassNotFoundException e){   out.print(e);    }
            StringBuffer str=new StringBuffer();
            Connection con;
            Statement sql;
            ResultSet rs;
            String uri="jdbc:odbc:bookshop";
            try{     con=DriverManager.getConnection(uri,"","");
                     sql=con.createStatement();
                     String s="SELECT * FROM user where logname='"+
                             logname+"'And password='"+oldPassword+"'" ;
                     rs=sql.executeQuery(s);
                     if(rs.next()){
                         String updateString="UPDATE user SET password='"+
                                    newPassword+"' where logname='"+logname+"'";
                         int m=sql.executeUpdate(updateString);
                         if(m==1) {
                             str.append("密码更新成功");
                         }
                         else{
                             str.append("密码更新失败");
                         }
                     }
                     else {
                         str.append("密码更新失败");
                     }
                     con.close();
            }
            catch(SQLException exp) {
                     str.append("密码更新失败"+exp);
            }
            jspContext.setAttribute("giveResult",str);
       %>
```

11.14 修改注册信息

该模块由一个 JSP 页面 ModifyRegister.jsp 和两个 Tag 文件 ModifyRegister.tag 和 GetRegister.tag 构成。ModifyRegister.jsp 页面负责调用 GetRegister.tag 文件，显示 GetRegister.tag 返回的用户曾注册的有关信息；ModifyRegister.jsp 页面调用 Tag 文件 ModifyRegister.tag 文件，并显示 ModifyRegister.tag 返回的有关修改注册信息是否成功的信息。

1．JSP 页面

ModifyRegister.jsp 页面调用 GetRegister.tag 文件，并显示 GetRegister.tag 返回的用户曾注册的有关信息。ModifyRegister.jsp 页面负责调用 Tag 文件 ModifyRegister.tag，并将用户名的新信息传递给该 Tag 文件。ModifyRegister.jsp 页面的效果如图 11.18 所示。

ModifyRegister.jsp.jsp
```
    <%@ page contentType="text/html;charset=GB2312" %>
```

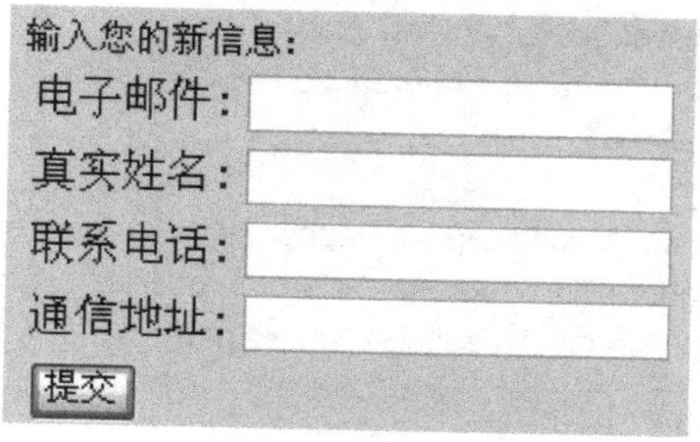

图 11.18 修改注册信息

```
<%@ taglib tagdir="/WEB-INF/tags" prefix="modifyRegister" %>
<%@ taglib tagdir="/WEB-INF/tags" prefix="getRegister" %>
<HEAD><%@ include file="head.txt" %></HEAD>
<HTML><BODY bgcolor=pink><CENTER>
<%
    boolean isModify=false;
    String logname=(String)session.getAttribute("logname");
    if(logname!=null){
        int m=logname.indexOf(",");
        logname=logname.substring(0,m);
        isModify=true;
    }
    else{
        response.sendRedirect("login.jsp");
    }
%>
<getRegister:GetRegister logname="<%=logname%>" />
<Font size=2>
<FORM action="" name=form>
<table>
    输入您的新信息:
    <tr><td>电子邮件:</td>
    <td><Input type=text name="email" value="<%=oldEmail%>"></td></tr>
    <tr><td>真实姓名:</td>
    <td><Input type=text name="realname" value="<%=oldRealname%>"></td></tr>
    <tr><td>联系电话:
    </td><td><Input type=text name="phone" value="<%=oldPhone%>"></td></tr>
    <tr><td>通信地址:
    </td><td><Input type=text name="address" value="<%=oldAddress%>"></td></tr>
    <tr><td><Input type=submit name="enter" value="提交"></td> </tr>
</table><Font></CENTER>
</BODY>
</HTML>
<%
    String enter=request.getParameter("enter");
    String email=request.getParameter("email");
    String realname=request.getParameter("realname");
```

```
            String phone=request.getParameter("phone");
            String address=request.getParameter("address");
            boolean ok=(enter!=null);
            if(ok&&isModify){
%>          <modifyRegister:ModifyRegister logname="<%=logname%>" email="<%=email%>"
            phone="<%=phone%>" address="<%=address%>" realname="<%=realname%>"/>
            <center><h2><%= giveResult %></h2></center>
<% }
%>
```

2．Tag 文件

GetRegister.tag 文件负责连接数据库，从 user 表查询用户曾注册的信息，ModifyRegister.tag 文件负责连接数据库，更新 user 表，以便改变用户的注册信息。

GetRegister.tag

```
<%@ tag import="java.sql.*" %>
<%@ tag pageEncoding="gb2312" %>
<%@ attribute name="logname" required="true" %>
<%@ variable name-given="oldEmail"   scope="AT_END" %>
<%@ variable name-given="oldAddress"   scope="AT_END" %>
<%@ variable name-given="oldRealname"   scope="AT_END" %>
<%@ variable name-given="oldPhone"   scope="AT_END" %>
<%
   try{ Class.forName("sun.jdbc.odbc.JdbcOdbcDriver");  }
   catch(ClassNotFoundException e){  out.print(e);  }
   StringBuffer str=new StringBuffer();
   Connection con;
   Statement sql;
   ResultSet rs;
   String uri="jdbc:odbc:bookshop";
   try{
          con=DriverManager.getConnection(uri,"","");
          String query=
          "select phone,email,address,realname WHERE logname='"+logname+"'";
          sql=con.createStatement();
          rs=sql.executeQuery(query);
          if(rs.next()){
             jspContext.setAttribute("oldPhone",rs.getString("phone"));
             jspContext.setAttribute("oldEmail",rs.getString("email"));
             jspContext.setAttribute("oldAddress",rs.getString("address"));
             jspContext.setAttribute("oldRealname",rs.getString("realname"));
          }
          else{
             jspContext.setAttribute("oldPhone","");
             jspContext.setAttribute("oldEmail","");
             jspContext.setAttribute("oldAddress","");
             jspContext.setAttribute("oldRealname","");
          }
```

```
            con.close();
        }
        catch(SQLException exp){
            jspContext.setAttribute("oldPhone","");
            jspContext.setAttribute("oldEmail","");
            jspContext.setAttribute("oldAddress","");
            jspContext.setAttribute("oldRealname","");
        }
%>
```

ModifyRegister.tag

```
<%@ tag import="java.sql.*" %>
<%@ tag pageEncoding="gb2312" %>
<%@ attribute name="logname" required="true" %>
<%@ attribute name="email" required="true" %>
<%@ attribute name="address" required="true" %>
<%@ attribute name="realname" required="true" %>
<%@ attribute name="phone" required="true" %>
<%@ variable name-given="giveResult" variable-class=
             "java.lang.StringBuffer" scope="AT_END" %>
<%
    byte [] c=email.getBytes("iso-8859-1");
    email=new String(c);
    c=address.getBytes("iso-8859-1");
    address=new String(c);
    c=realname.getBytes("iso-8859-1");
    realname=new String(c);
    c=phone.getBytes("iso-8859-1");
    phone=new String(c);
    try{  Class.forName("sun.jdbc.odbc.JdbcOdbcDriver");  }
    catch(ClassNotFoundException e){  out.print(e);  }
    StringBuffer str=new StringBuffer();
    Connection con;
    Statement sql;
    ResultSet rs;
    String uri="jdbc:odbc:bookshop";
    try{
            con=DriverManager.getConnection(uri,"","");
            String updateCondition="UPDATE user SET phone='"+
                phone+"',email='"+email+"',address='"+
                address+"',realname='"+
                realname+"'   WHERE logname='"+logname+"'";
            sql=con.createStatement();
            int m=sql.executeUpdate(updateCondition);
            if(m==1) {
                str.append("修改信息成功");
            }
            else {
```

```
                    str.append("更新失败");
                }
                con.close();
        }
        catch(SQLException exp){
            str.append("更新失败"+exp);
        }
        jspContext.setAttribute("giveResult",str);
%>
```

11.15 退出登录

该模块只有一个名字为 ExitLogin.jsp 的页面，负责销毁用户的 session 对象，导致登录失效。ExitLogin.jsp 页面的效果如图 11.19 所示。

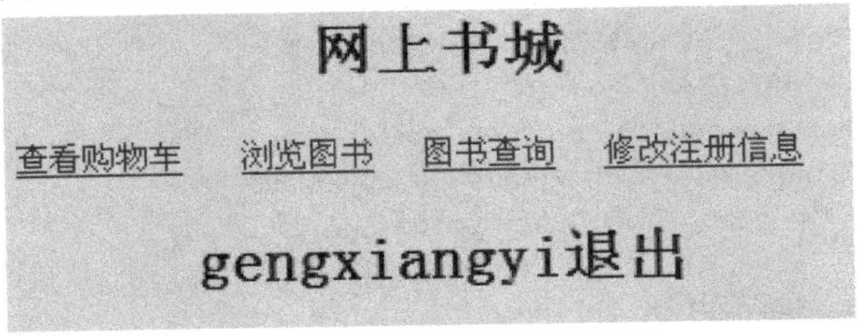

图 11.19 退出登录

ExitLogin.jsp

```
<%@ page contentType="text/html;charset=GB2312" %>
<HEAD><%@ include file="head.txt" %></HEAD>
<HTML><BODY bgcolor=pink><CENTER>
<%
    String logname=(String)session.getAttribute("logname");
    if(logname!=null){
        int m=logname.indexOf(",");
        logname=logname.substring(0,m);
        session.invalidate();
        out.print("<h2>"+logname+"退出</h2>");
    }
    else{
        response.sendRedirect("login.jsp");
    }
%>
```